**Also by Marvin Minsky**

THE SOCIETY OF MIND

# THE
# EMOTION
# MACHINE

Commonsense Thinking,
Artificial Intelligence, and the
Future of the Human Mind

# MARVIN MINSKY

Simon & Schuster

New York   London   Toronto   Sydney

SIMON & SCHUSTER
Rockefeller Center
1230 Avenue of the Americas
New York, NY 10020

Verse on page 298 reprinted courtesy of Theodore Melnechuk

For information regarding special discounts for bulk purchases,
please contact Simon & Schuster Special Sales:
1-800-456-6798 or business@simonandschuster.com.

Designed by Nancy Singer Olaguera

Manufactured in the United States of America

10  9  8  7  6  5  4  3  2

Library of Congress Cataloging-in-Publication Data

Minsky, Marvin Lee.
    The emotion machine : commonsense thinking,
    artificial intelligence, and the future of the human
    mind / Marvin Minsky.
        p. cm.
Includes bibliographical references (p.      ) and index.
1. Human information processing. 2. Emotions and cognition. I. Title.

BF444.M56  2006
153—dc22                                              2006044367

ISBN-13: 978-0-7432-7663-4
ISBN-10:      0-7432-7663-9

# CONTENTS

# THE
# EMOTION
# MACHINE

# INTRODUCTION

*Nora Joyce, to her husband James:*
*"Why don't you write books people can read?"*

I hope this book will be useful to everyone who seeks ideas about how human minds might work, or who wants suggestions about better ways to think, or who aims toward building smarter machines. It should be useful to readers who want to learn about the field of Artificial Intelligence. It should also be of interest to psychologists, neurologists, computer scientists, and philosophers because it develops many new ideas about the subjects those specialists struggle with.

We all admire great accomplishments in the sciences, arts, and humanities—but we rarely acknowledge how much we achieve in the course of our everyday lives. We recognize the things we see, we understand the words we hear, and we remember things that we've experienced so that, later, we can apply what we've learned to other kinds of problems and opportunities.

We also do a remarkable thing that no other creatures seem able to do: whenever our usual ways to think fail, *we can start to think about our thoughts themselves*—and if this "reflective thinking" shows where we went wrong, that can help us to invent new and more powerful ways to think. However, we still know very little about how our brains manage to do such things. How does imagination work? What are the causes of consciousness? What are emotions, feelings, and thoughts? How do we manage to think at all?

Contrast this with the progress we've seen toward answering questions about physical things. What are solids, liquids, and gases? What are colors, sounds, and temperatures? What are forces, stresses, and strains? What is the nature of energy? Today, almost all such mysteries have been explained in terms of very small numbers of simple laws—for example, the equations discovered by such physicists as Newton, Maxwell, Einstein, and Schrödinger.

So naturally, psychologists tried to imitate physicists—by searching for compact sets of laws to explain what happens inside our brains. However, no such simple set of laws exists, because every brain has hundreds of parts, each of which evolved to do certain particular kinds of jobs; some of them recognize situations, others tell muscles to execute actions, others formulate goals and plans, and yet others accumulate and use enormous bodies of knowledge. And though we don't yet know enough about how each of those brain-centers works, we do know their construction is based on information that is contained in tens of thousands of inherited genes, so that each brain-part works in a way that depends on a somewhat different set of laws.

Once we recognize that our brains contain such complicated machinery, this suggests that we need to do the opposite of what those physicists did: instead of searching for simple explanations, we need to find more complicated ways to explain our most familiar mental events. The meanings of words like "feelings," "emotions," or "consciousness" seem so natural, clear, and direct to us that we cannot see how to start thinking about them. However, this book will argue that none of those popular psychology words refers to any single, definite process; instead each of those words attempts to describe the effects of large networks of processes inside our brains. For example, Chapter 4 will demonstrate that "consciousness" refers to more than twenty different such processes!

It might appear to make everything worse, to change some things that looked simple at first into problems that now seem more difficult. However, on a larger scale, this increase in complexity will actually make our job easier. For, once we split each old mystery into parts, we will have replaced each old, big problem with several new and smaller ones—each of which may still be hard but no longer will seem unsolvable. Furthermore, Chapter 9 will argue that regarding ourselves as complex machines need not diminish our feelings of self-respect, and should enhance our sense of responsibility.

To start dividing those old big questions into smaller ones, this book will begin by portraying a typical brain as containing a great many parts that we'll call "resources."

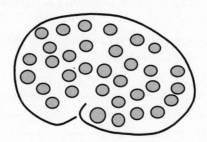

We'll use this image whenever we want to explain some mental activity (such as Anger, Love, or Embarrassment) by trying to show how that state of mind might result from the activities of a certain collection of mental resources. For example, the state called "Anger" appears to arouse resources that make us react with unusual speed and strength—while suppressing resources that we otherwise use to plan and act more prudently; thus, Anger replaces your cautiousness with aggressiveness and trades your sympathy for hostility. Similarly, the condition called "Fear" would engage resources in ways that cause you to retreat.

> Citizen: I sometimes find myself in a state where everything seems cheerful and bright. Other times (although nothing has changed) all my surroundings seem dreary and dark, and my friends describe me as "down" or "depressed." Why do I have such states of mind—or moods, or feelings, or dispositions—and what causes all of their strange effects?

Some popular answers to this are, *"Those changes are caused by chemicals in the brain,"* or *"They result from an excess of stress,"* or *"They come from thinking depressing thoughts."* However, such statements say almost nothing about how those processes actually work—whereas the idea of selecting a set of resources can suggest more specific ways in which our thinking can change. For example, Chapter 1 will begin by thinking about this very familiar phenomenon:

When a person you know has fallen in love, it's almost as though someone new has emerged—a person who thinks in other ways, with altered goals and priorities. It's almost as though a switch had been thrown and a different program has started to run.

What could happen inside a brain to make such changes in how it thinks? Here is the approach this book will take:

> *Each of our major "emotional states" results from turning certain resources on while turning certain others off—and thus changing some ways that our brains behave.*

But what activates such sets of resources? Our later chapters will argue that our brains must also be equipped with resources that we shall call *"Critics"*—each of which is specialized to recognize some certain condition—and then to activate a specific collection of other resources. Some of our Critics are built in from birth, to provide us with certain "instinctive" reactions—such as Anger, Hunger, Fear, and Thirst—which evolved to help our ancestors survive. Thus, Anger and Fear evolved for defense and protection, while Hunger and Thirst evolved for nutrition.

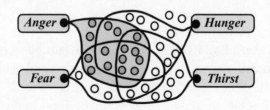

However, as we learn and grow, we also develop ways to activate other, new sets of resources to use—and this leads to types of mental states that we regard as more "intellectual" than "emotional." For example, whenever a problem seems hard to you, then your mind will start to switch among different Ways to Think—by selecting different sets of resources that can help you to divide the problem into smaller parts, or find suggestive analogies, or retrieve solutions from memories—or even ask some other person for help.

The rest of this book will argue that this could be what provides our species with our uniquely human resourcefulness.

*Each of our major Ways to Think results from turning certain resources on while turning certain others off—and thus changing some ways that our brains behave.*

For example, our first few chapters will try to show how this could explain such states of mind as Love, Attachment, Grief, and Depression in terms of how they exploit our resources. Then the later chapters will do the same for more "intellectual" sorts of thought.

Citizen: It seems strange that you've given the same description both for emotions and for regular thinking. But thinking is basically rational—dry, detached, and logical—whereas emotions enliven our ways to think by adding irrational feelings and biases.

There is a traditional view in which emotions *add* extra features to plain, simple thoughts, much as artists use colors to augment the effects of black-and-white drawings. However, this book will argue, instead, that many of our emotional states result when certain particular Ways to Think start to *suppress* our use of certain resources! For example, Chapter 1 will portray "infatuation" as a condition in which we suppress some resources that we might otherwise use to recognize faults in somebody else. Besides, I think it's a myth that there's any such thing as purely logical, rational thinking—because our minds are always affected by our assumptions, values, and purposes.

Citizen: I still think your view of emotions ignores too much. For example, emotional states like fear and disgust involve the body as well as the brain, as when we feel discomfort in the chest or gut, or palpitations of the heart, or when we feel faint or tremble or sweat.

I agree that this view may seem too extreme—but sometimes, to explore new ideas, we need to set our old ones aside, at least temporarily. For example, in the most popular view, emotions are deeply involved with our bodies' conditions. However, Chapter 7 will take the opposite view, by regarding our body parts as resources that our brains can use to change (or maintain) their mental states! For example, you sometimes can make yourself persist at a plan by maintaining a certain facial expression.

So, although this book is called *The Emotion Machine,* it will argue that emotional states are not especially different from the processes that we call "thinking"; instead, emotions are certain ways to think that we use to increase our resourcefulness—that is, when our passions don't grow till they handicap us—and this variety of Ways to Think must be such a substantial part of what we call "intelligence" that perhaps we should call it "resourcefulness." And this applies not only to emotional states but to all of our mental activities:

> *If you "understand"' something in only one way, then you scarcely understand it at all—because when you get stuck, you'll have nowhere to go. But if you represent something in several ways, then when you get frustrated enough, you can switch among different points of view, until you find one that works for you!*

Accordingly, when we design machines to mimic our minds—that is, to create Artificial Intelligences—we'll need to make sure that those machines, too, are equipped with sufficient diversity:

> If a program works in only one way, then it gets stuck when that method fails. But a program that has several ways to proceed could then switch to some other approach, or search for a suitable substitute.

This idea is a central theme of this book—and it is firmly opposed to the popular view that each person has a central core—some sort of invisible spirit or self—from which all their mental abilities originate. For that seems a demeaning idea—that all our virtues are secondhand—or that we deserve no credit for our accomplishments, because they come to us as gifts from some other source. Instead, I see our dignity as stemming from what we each have made of ourselves: a colossal collection of different ways to deal with different situations and predicaments. It is that diversity that distinguishes us from most of the other animals—and from all the machines that we've built in the past—and every chapter of this book will discuss some of the sources of our uniquely human resourcefulness.

For centuries, psychologists searched for ways to explain our everyday mental processes—yet many thinkers still today regard the nature of mind as a mystery. Indeed, it still is widely believed that minds are made of ingredients that can only exist in living things, that no machine could feel or think, worry about what might happen to it, or even be conscious that it exists—or could ever develop the kinds of ideas that could lead to great paintings or symphonies.

This book will pursue all those goals at once: to suggest how human brains might work and to design machines that can feel and think. Then we can try to apply those ideas both to understand ourselves and to develop Artificial Intelligence.

## How This Book Handles Quotations and References

Each statement in quotation marks is by an actual person; if it also has a date, the source will be in the bibliography.

> Marcel Proust 1927: "Each reader reads only what is already inside himself. A book is only a sort of optical instrument which the writer offers to let the reader discover in himself what he would not have found without the aid of the book."

A statement without quotation marks is a fictional comment a reader might make.

> Citizen: If our everyday thinking is so complex, then why does it seem so straightforward to us?

Most references are conventional bibliographical citations, such as

> Schank, 1975: Roger C. Schank, *Conceptual Information Processing*. New York: American Elsevier, 1975.

Some references are to pages on the World Wide Web.

> Lenat 1998: Douglas B. Lenat. *The Dimensions of Context Space*. Available at http://www.cyc.com/doc/context-space.pdf.

Some other references are to "newsgroups" on the Web, such as

> McDermott 1992: Drew McDermott, In comp.ai.philosophy. February 7, 1992.

To access such newsgroup documents (along with the context in which they were written) one can make a Google search for comp.ai.philosophy McDermott 1992. So I will try to maintain copies of these on my Web site at www.emotionmachine.net. Readers are also invited to use that site for sending questions and comments to me.

*Note:* This book uses the term *"resource"* where my earlier book, *The Society of Mind,* used *"agent."* I made this change because too many readers assumed that an "agent" is a personlike thing (like a travel agent) that could operate independently, or cooperate with others in much the same ways that people do. On the contrary, most resources are specialized to certain kinds of jobs for certain other resources, and cannot directly communicate with most of the person's other resources. For more details about how these two books relate, see the article by Push Singh 2003, who helped to develop many of the ideas in this book.

# 1
# FALLING IN LOVE

## 1-1 Infatuation

> "In faith, I do not love thee with mine eyes,
> For they in thee a thousand errors note;
> But 'tis my heart that loves what they despise."
> —*Shakespeare*

Many people find it absurd to think of a person as like a machine—so we often hear such statements as this:

Citizen: Of course machines can do useful things. We can make them add up huge columns of numbers or assemble cars in factories. But nothing made of mechanical stuff could ever have genuine feelings like love.

No one finds it surprising these days when we make machines that do logical things, because logic is based on clear, simple rules of the sorts that computers can easily use. But *Love,* by its nature, some people would say, cannot be explained in mechanical ways—nor could we ever make machines that possess any such human capacities as feelings, emotions, and consciousness.

What is Love, and how does it work? Is this something that we want to understand, or is it one of those subjects that we don't really want to know more about? Hear our friend Charles attempt to describe his latest infatuation.

"I've just fallen in love with a wonderful person. I scarcely can think about anything else. My sweetheart is unbelievably perfect—of indescribable beauty, flawless character, and incredible intelligence. There is nothing I would not do for her."

On the surface such statements seem positive; they're all composed of superlatives. But note that there's something strange about this: most of those phrases of positive praise use syllables like "*un,*" "*less,*" and "*in*"—which show that they really are negative statements describing the person who's saying them!

Wonderful. Indescribable.
    (I can't figure out what attracts me to her.)
I scarcely can think of anything else.
    (Most of my mind has stopped working.)
Unbelievably perfect. Incredible.
    (No sensible person believes such things.)
She has a flawless character.
    (I've abandoned my critical faculties.)
There is nothing I would not do for her.
    (I've forsaken most of my usual goals.)

Our friend sees all this as positive. It makes him feel happy and more productive, and relieves his dejection and loneliness. But what if most of those pleasant effects result from his success at suppressing his thoughts about what his sweetheart actually says:

"Oh, Charles—a woman needs certain things. She needs to be loved, wanted, cherished, sought after, wooed, flattered, cosseted, pampered. She needs sympathy, affection, devotion, understanding, tenderness, infatuation, adulation, idolatry—that isn't much to ask, is it, Charles?"[1]

Thus, Love can make us disregard most defects and deficiencies, and make us deal with blemishes as though they were embellishments—even when, as Shakespeare said, we still may be partly aware of them:

"When my love swears that she is made of truth,
I do believe her, though I know she lies."

We are equally apt to deceive ourselves, not only in our personal lives but also when dealing with abstract ideas. There, too, we often close our eyes to conflicts and clashes between our beliefs. Listen to Richard Feynman's words:

> "That was the beginning and the idea seemed so obvious to me that I fell deeply in love with it. And, like falling in love with a woman, it is only possible if you don't know too much about her, so you cannot see her faults. The faults will become apparent later, but after the love is strong enough to hold you to her. So, I was held to this theory, in spite of all the difficulties, by my youthful enthusiasm."
> —*1966 Nobel Prize lecture*

What does a lover actually love? That should be the person to whom you're attached—but if your pleasure mainly results from suppressing your other questions and doubts, then you're only in love with Love itself.

> Citizen: So far, you have spoken only about what we call infatuation—sexual lust and extravagant passion. That leaves out most of the usual meanings of "love"—such as tenderness, trust, and companionship.

Indeed, once those short-lived attractions fade, they sometimes go on to be replaced by more enduring relationships, in which we exchange our own interests for those of the persons to whom we're attached:

> Love, n. That disposition or state of feeling with regard to a person which (arising from recognition of attractive qualities, from instincts of natural relationship, or from sympathy) manifests itself in solicitude for the welfare of the object, and usually also in delight in his or her presence and desire for his or her approval; warm affection, attachment.
> —Oxford English Dictionary

Yet even this larger conception of Love is still too narrow to cover enough, because *Love* is a kind of suitcase-like word, which includes other kinds of attachments like these:

The love of a parent for a child.
A child's affection for parents and friends.
The bonds that make lifelong companionships.
The connections of members to groups or their leaders.

We also apply that same word Love to our involvements with objects, feelings, ideas, and beliefs—and not only to ones that are sudden and brief, but also to bonds that increase through the years.

A convert's adherence to doctrine or scripture.
A patriot's allegiance to country or nation.
A scientist's passion for finding new truths.
A mathematician's devotion to proofs.

Why do we pack such dissimilar things into those single suitcase-words? As we'll see in Section 1-3, each of our common "emotional" terms describes a variety of different processes. Thus we use the word Anger to abbreviate a diverse collection of mental states, some of which change our ways to perceive, so that innocent gestures get turned into threats—and thus make us more inclined to attack. Fear also affects the ways we react but makes us retreat from dangerous things (as well as from some that might please us too much).

Returning to the meanings of Love, one thing seems common to all those conditions: *each leads us to think in different ways:*

*When a person you know has fallen in love, it's almost as though someone new has emerged—a person who thinks in other ways, with altered goals and priorities. It's almost as though a switch had been thrown and a different program has started to run.*

This book is mainly filled with ideas about what could happen inside our brains to cause such great changes in how we think.

## 1-2 The Sea of Mental Mysteries

From time to time we think about how we try to manage our minds:

Why do I waste so much of my time?
What determines to whom I'm attracted?

Why do I have such strange fantasies?
Why do I find mathematics so hard?
Why am I afraid of heights and crowds?
What makes me addicted to exercise?

But we can't hope to understand such things without adequate answers to questions like these:

What sorts of things are emotions and thoughts?
How do our minds build new ideas?
What are the bases for our beliefs?
How do we learn from experience?
How do we manage to reason and think?

In short, we all need better ideas about the ways in which we think. But whenever we start to think about that, we encounter yet more mysteries.

What is the nature of consciousness?
What are feelings and how do they work?
How do our brains imagine things?
How do our bodies relate to our minds?
What forms our values, goals, and ideals?

Now, everyone knows how Anger feels—or Pleasure, Sorrow, Joy, and Grief—yet we still know almost nothing about how those processes actually work. As Alexander Pope asks in his *Essay on Man,* are these things that we can hope to understand?

"Could he, whose rules the rapid comet bind,
Describe or fix one movement of his mind?
Who saw its fires here rise, and there descend,
Explain his own beginning, or his end?"

How did we manage to find out so much about atoms and oceans and planets and stars—yet so little about the mechanics of minds? Thus, Newton discovered just three simple laws that described the motions of all sorts of objects; Maxwell uncovered just four more laws that explained all electromagnetic events; then Einstein reduced all those and more into yet smaller formulas. All this came from the success of those physicists' quest: *to find simple explanations for things that seemed, at first, to be highly complex.*

Then, why did the sciences of the mind make less progress in those same three centuries? I suspect that this was largely because most psychologists mimicked those physicists, by looking for equally compact solutions to questions about mental processes. However, that strategy never found small sets of laws that accounted for, in substantial detail, any large realms of human thought. So this book will embark on the opposite quest: *to find* more complex *ways to depict mental events that seem simple at first!*

This policy may seem absurd to scientists who have been trained to believe such statements as, *"One should never adopt hypotheses that make more assumptions than one needs."* But it is worse to do the opposite—as when we use "psychology words" that mainly hide what they try to describe. Thus, every phrase in the sentence below conceals its subject's complexities:

*You look at an object and see what it is.*

For, *"look at"* suppresses your questions about the systems that choose how you move your eyes. Then, *"object"* diverts you from asking how your visual systems partition a scene into various patches of color and texture—and then assign them to different "things." Similarly, *"see what it is"* serves to keep you from asking how recognitions relate to other things that you've seen in the past.

It is the same for most of the commonsense words we use when we try to describe the events in minds—as when one makes a statement like, *"I think I understood what you said."* Perhaps the most extreme examples of this are when we use words like *you* and *me,* because we all grow up with this fairy tale:

*We each are constantly being controlled by powerful creatures inside our minds who do our feeling and thinking for us, and make our important decisions for us. We call these our "Selves" or "Identities"— and we believe that they always remain the same, no matter how we may otherwise change.*

This *"Single-Self"* concept serves us well in our everyday social affairs. But it hinders our efforts to think about what minds are and how they work—because, when we ask about what Selves actually do, we get the same answer to every such question:

*Your Self sees the world by using your senses. Then it stores what it learns in your memory. It originates all your desires and goals— and then solves all your problems for you, by exploiting your "intelligence."*

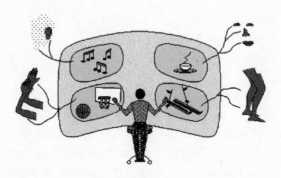

A SELF CONTROLLING ITS PERSON'S MIND

What attracts us to this queer idea, that we don't make any decisions ourselves but delegate them to some other entity? Here are a few kinds of reasons why a mind might entertain such a fiction:

Child psychologist: As a child, you learned to distinguish among some persons in your environment. Later, you somehow came to conclude that you are such a person, too—but at the same time, you may have assumed that there is a person inside of you.

Psychotherapist: The Single-Self legend helps makes life seem pleasant, by hiding from us how much we're controlled by all sorts of conflicting, unconscious goals.

Practical person: That image makes us efficient, whereas better ideas might slow us down. It would take too long for our hard-working minds to understand everything all the time.

However, although the Single-Self concept has practical uses, it does not help us to understand ourselves—because it does not provide us with *smaller parts* we could use to build theories of what we are. When you think of yourself as a single thing, this gives you no clues about issues like these:

What determines the subjects I think about?
How do I choose what next to do?
How can I solve this difficult problem?

Instead, the Single-Self concept offers only useless answers like these:

My Self selects what to think about.
My Self decides what I should do next.
I should try to make my Self get to work.

Whenever we wonder about our minds, the simpler are the questions we ask, the harder it seems to find answers to them. When asked about a complex physical task like, *"How could a person build a house,"* you might answer almost instantly, *"Make a foundation and then build walls and a roof."* However, we find it much harder to think of what to say about seemingly simpler questions like these:

How do you recognize things that you see?
How do you comprehend what a word means?
What makes you like pleasure more than pain?

Of course, those questions are not really simple at all. To "see" an object or "speak" a word involves hundreds of different parts of your brain, each of which does some quite difficult jobs. Then why don't we sense that complexity? That's because most such jobs are done inside parts of the brain whose internal processes are hidden from the rest of the brain.

At the end of this book, we'll come back to examine the concepts of Self and Identity, and conclude that the structures that we call our Selves are elaborate structures that each of us builds to use for many purposes.

*Whenever you think about your "Self," you are switching among a huge network of models, each of which tries to represent some particular aspects of your mind—to answer some questions about yourself.*

## 1-3 Moods and Emotions

William James 1890: "If one should seek to name each particular one of them of which the human heart is the seat, each race of men having found names for some shade of feeling which other races have left undiscriminated . . . all sorts of groupings would be possible, according as we chose this character or that as a basis. The only question would be, does this grouping or that suit our purpose best?"

Sometimes a person gets into a state where everything seems to be cheerful and bright—although nothing outside has actually changed. Other times everything pleases you less: the entire world seems dreary and dark, and your friends complain that you seem depressed. Why do we have such states of mind—or moods, or feelings, or dispositions—and what causes all their strange effects? Here are some of the phrases we find when dictionaries define *emotion*.

The subjective experience of a strong feeling.
A state of mental agitation or disturbance.
A mental reaction involving the state of one's body.
A subjective rather than conscious affection.
The parts of consciousness that involve feeling.
A nonrational aspect of reasoning.

If you didn't yet know what emotions are, you certainly wouldn't learn much from this. What is *subjective* supposed to mean, and what could a *conscious affection* be? In what ways do those *parts of consciousness* become *involved* with what we call *"feelings"*? Must every emotion involve a *disturbance*? Why do so many such questions arise when we try to define what *emotion* means?

The reason for this is simply that *emotion* is one of those suitcaselike words that we use to conceal the complexity of very large ranges of different things whose relationships we don't yet comprehend. Here are a few of the hundreds of terms that we use to refer to our mental conditions:

Admiration, Affection, Aggression, Agitation, Agony, Alarm, Ambition, Amusement, Anger, Anguish, Anxiety, Apathy, Assurance, Attraction, Aversion, Awe, Bliss, Boldness, Boredom, Confidence, Confusion, Craving, Credulity, Curiosity, Dejection, Delight, Depression, Derision, Desire, Detest, Disgust, Dismay, Distrust, Doubt, etc.

Whenever you change your mental state, you might try to use those emotion-words to try to describe your new condition—but usually each such word or phrase refers to too wide a range of states. Many researchers have spent their lives at classifying our states of mind, by arranging terms like *feelings, dispositions, tempers,* and *moods* into orderly charts or diagrams—but should we call *Anguish* a feeling or a mood? Is *Sorrow* a type of disposition? No one can settle the use of such terms because different traditions make different distinctions, and different people have different ideas about how to describe their various states of mind. How many readers can claim to know precisely how each of the following feelings feels?[2]

> Grieving for a lost child
> Fearing that nations will never live in peace
> Rejoicing in an election victory
> Excited anticipation of a loved one's arrival
> Terror as your car loses control at high speed
> Joy at watching a child at play
> Panic at being in an enclosed space

In everyday life, we expect our friends to know what we mean by *Pleasure* or *Fear*—but I suspect that attempting to make our old words more precise has hindered more than helped us to make theories about how human minds work. So this book will take a different approach, by thinking of each mental condition as based on the use of many small processes.

## 1-4  Infant Emotions

> Charles Darwin 1872: "Infants, when suffering even slight pain, moderate hunger, or discomfort, utter violent and prolonged screams. Whilst thus screaming their eyes are firmly closed, so that the skin round them is wrinkled, and the forehead contracted into a frown. The mouth is widely opened with the lips retracted in a peculiar manner, which causes it to assume a squarish form; the gums or teeth being more or less exposed."

One moment your baby seems perfectly well, but then come some restless motions of limbs. Next you see a few catches of breath, and then suddenly the air fills with screams. Is baby hungry, sleepy, or wet? Whatever the trouble may turn out to be, those cries compel you to find some way to help—and once you find the remedy, things quickly return to normal. In the meantime though, you, too, feel distressed. When a friend of yours cries, you can ask her what's wrong—but when your baby abruptly changes his state, there may seem to be "no one home" to communicate with.

Of course, I do not mean to suggest that infants don't have "personalities." Soon after birth you can usually sense that a particular baby reacts more quickly than others, or seems more patient or irritable, or even more inquisitive. Some of those traits may change with time, but others persist throughout life. Nevertheless, we still need to ask, What could make an infant so suddenly switch, between one moment and the next, from contentment or calmness to anger or rage?

To answer that kind of question, you would need a theory about the machinery that underlies that infant's behavior. So let's imagine that someone has asked you to build an artificial animal. You could start by making a list of goals that your animal-robot needs to achieve. It may need to find parts with which to repair itself. It may need defenses against attacks. Perhaps it should regulate its temperature. It may even need ways to attract helpful friends. Then once you have assembled that list, you could tell your engineers to meet each of those needs by building a separate "instinct-machine"—and then to package them all into a single "body-box."

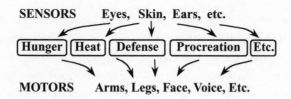

What goes inside each instinct-machine? Each of them needs three kinds of resources: some ways to recognize situations, some knowledge about how to react to these, and some muscles or motors to execute actions.

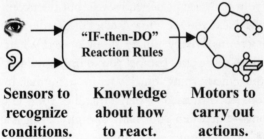

**Sensors to recognize conditions.** **Knowledge about how to react.** **Motors to carry out actions.**

What goes inside that knowledge box? Let's begin with the simplest case: suppose that we already know, in advance, all the situations our robot will face. Then all we need is a catalog of simple, two-part *"If→Do"* rules— where each *If* describes one of those situations, and each *Do* describes an action to take. Let's call this a *"Rule-Based Reaction-Machine."*

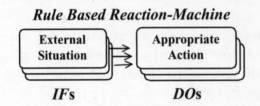

*If* you are too hot, *Move* into the shade.
*If* you are hungry, *Find* something to eat.
*If* you're facing a threat, *Select* some defense.

Every infant animal is born with many *If→Do* rules like these. For example, each human infant is born with ways to maintain its body temperature: when too hot, it can pant, sweat, stretch out, or vasodilate; when too cold, it can shiver, retract its limbs, or vasoconstrict—or metabolize to produce more heat. Then later in life, we learn to use actions that change the external world.

*If* you are too cold, *Turn* on a heater.
*If* your room is too hot, *Open* a window.
*If* there's too much sunlight, *Pull* down the shade.

It would be naive to try to describe a mind as nothing more than bundles of *If→Do* rules. However, the great animal psychologist Nikolaas Tinbergen showed in his book *The Study of Instinct*[3] that when such rules are combined in certain ways, they can account for a remarkable range of different things that animals do. This sketch shows only a part of the structure that Tinbergen proposed to explain how a certain fish behaves.

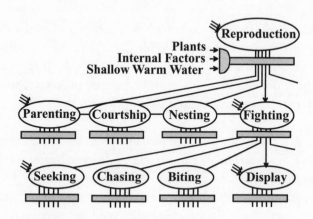

Of course, it would need much more than this to support the higher levels of human thought. The rest of this book will describe some ideas about the structures inside our human minds.

## 1-5 Seeing a Mind As a Cloud of Resources

We all know ways to describe our minds, as they appear to us when seen from outside:

> Albert Einstein 1950: "We are all ruled in what we do by impulses; and these impulses are so organized that our actions in general serve for our self preservation and that of the race. Hunger, love, pain, fear are some of those inner forces which rule the individual's instinct for self preservation. At the same time, as social beings, we are moved in the relations with our fellow beings by such feelings as sympathy, pride, hate, need for power, pity, and so on."

This book will try to show how such states of mind could come from machines inside our brains. To be sure, many thinkers still insist that machines can never feel or think.

Citizen: A machine can do only what it is programmed to do, and does it without any thinking or feeling. No machine can get tired or bored or have any kind of emotion at all. It cannot care when something goes wrong, and, even when it gets things right, it feels no sense of pleasure, pride, or delight in those accomplishments.

Vitalist: That's because machines have no spirits or souls, and no wishes, ambitions, desires, or goals. That's why a machine will just stop when it's stuck—whereas a person will struggle to get something done. Surely this must be because people are made of different stuff; we are alive and machines are not.

In earlier times, those views seemed plausible, because living things seemed so different from machines—and no one could even begin to conceive of how physical things could feel or think. But once we developed more scientific instruments (and better ideas about science itself), then "life" became less mysterious, because now we could see that each living cell consists of hundreds of kinds of machinery.

Holist: Yes, but many people still maintain that there will always remain a mystery about how a living thing could ever result from nothing more than mechanical stuff. Surely we're more than the sum of our parts.

That once was a popular belief, but today it is widely recognized that behavior of a complex machine depends only on how its parts interact, but not on the "stuff" of which they are made (except for matters of speed and strength). In other words, all that matters is the manner in which each part reacts to the other parts to which it is connected. For example, we can build computers that behave in identical ways, no matter if they consist of electronic chips or of wood and paper clips—provided that their parts perform the same processes, so far as the other parts can see.

This suggests replacing old questions like, "What sorts of things are emotions and thoughts?" by more constructive ones like, "What *processes* does each emotion involve?" and "How could machines perform such *processes*?" To do this, we'll start with the simple idea that every brain contains many parts, each of which does certain specialized jobs. Some can recognize various patterns, others can supervise various actions, yet others can formulate goals or plans, and some can contain large bodies of knowledge. This suggests that we could envision a mind (or a brain) as composed of a great many different "resources."

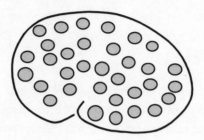

At first this image may seem hopelessly vague—yet it can help us start to understand how a mind could make a large change in its state. For example, the state we call "angry" could be what happens when you activate some resources that help you react with more speed and strength—while also suppressing some other resources that usually make you act prudently. This will replace your usual cautiousness with aggressiveness, change empathy into hostility, and cause you to plan less carefully. All of this could result from turning on the resource labeled Anger in this diagram:

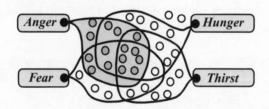

Similarly, we could explain such mental conditions as Hunger and Fear—and we could even account for what happened to Charles in his state of acute infatuation: perhaps such a process turned off the resources he normally used to recognize another person's faults—and also supplanted some of his usual goals by ones that he thought Celia wants him to hold. So now, let's make a generalization:

*Each of our major "emotional states" results from turning certain resources on while turning certain others off—thus changing the way one's brain behaves.*

And although that may seem like an oversimplification, we'll take it to a further extreme, because we see emotional states as particular types of Ways to Think.

> *Each of our various Ways to Think results from turning certain resources on while turning certain others off—thus changing the way one's brain behaves.*

In this way, we can regard our mental states as what happens when different sets of resources interact, and most of this book will be about how some of those mental resources might work. First, perhaps, we ought to ask how those resources originate. Clearly, some of them must have evolved to promote functions that keep our bodies alive; Anger and Fear evolved for protection, and Hunger evolved to serve nutrition—and many such "basic instincts" are already built into our brains at birth. Other resources appear in later years, such as the ones involved with reproduction (which often engages some risky behaviors); some of these also must be inborn, but others must be mainly learned.

What happens when several selections are turned on at once, so that some resources get both aroused and suppressed? This could lead to some of the mental states in which we say, "Our feelings are mixed." For example, when one detects some sort of threat, this might arouse parts of both Anger and Fear.

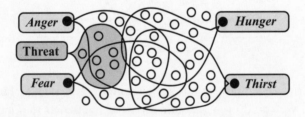

Then if one tried both to attack and retreat, that could lead to paralysis—and that sometimes occurs in some animals. However, human minds can escape from such traps, as we'll see in some later chapters, by using "higher-level" resources to help to settle such conflicts.

Student: I could better grasp what you're talking about if you could be a bit more precise about what you mean by the word *resource*. Do you imagine that each resource has a separate, definite place in the brain?

I'm using *resource* in a hazy way, to refer to all sorts of structures and processes that range from perception and action to Ways to Think about bodies of knowledge. Some such functions are performed in certain particular parts of the brain, while others use parts that are more widely spread over much larger portions of the brain. Other parts of this book discuss more ideas about the kinds of resources our brains seem to support, as well as how their functions might be organized. However, I won't try to identify where these might lie in the brain because research on this is advancing so quickly that any conclusion one might make today could be outdated in just a few weeks.

As we said, this resource-cloud idea may at first seem too vague—but as we develop more detailed ideas about how our mental resources behave, we'll gradually replace it with more elaborate theories about how our mental resources are organized.

> Student: You speak of a person's emotional states as nothing more than ways to think, but surely that's too cold and abstract—too intellectual, dull, and mechanical. Besides, it doesn't explain the pleasures and pains that come when we succeed or fail, or the thrills that we experience from works of artistic genius.

> Rebecca West: "It overflows the confines of the mind and becomes an important physical event. The blood leaves the hands, the feet, the limbs, and flows back to the heart, which for the time seems to have become an immensely high temple whose pillars are several sorts of illumination, returning to the numb flesh diluted with some substance swifter and lighter and more electric than itself."[4]

Many traditional views of emotions emphasize the extent to which events that occur in our body parts can affect our mental processes—as when we experience muscular tensions. However, our brains do not directly detect those tensions, but only react to signals that come through nerves that connect to those body parts. So while our bodies can play important roles, we can also regard our bodies, too, as composed of resources our brains can exploit.

The rest of this book will focus on what sort of mental resources we have, what kinds of things each resource might do, and how each affects those to which it is connected. We'll begin by developing more ideas about what turns resources off and on.

Student: Why should one ever turn off a resource? Why not keep them all working all the time?

Indeed, certain resources are *never* switched off—such as those involved with vital functions like respiration, balance, and posture—or those that constantly keep watch for certain particular types of danger. However, if *all* our resources were active at once, they would too often get into conflicts. You can't make your body both walk and run, or move in two different directions at once. So when one has several goals that are incompatible, because they compete for the same resources (or for time, space, or energy), then one needs to engage processes that have ways to manage such conflicts.

It is much the same in a human society: when different people have different goals, they may be able to pursue these separately. But when this leads to excessive conflict or waste, societies often then create multiple levels of management in which (at least in principle) each manager controls the activities of certain lower-level individuals.

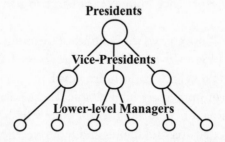

However, both in societies and in brains, few "higher-level executives" know enough of the system's details to specify what must be done—hence, much of their "power" in fact consists in selecting among options proposed by their subordinates. Then, in effect, those low-level individuals will, at least transiently, be controlling or constraining what their superiors do.

For example, whenever some mental process gets stuck, it may need to split the problem into smaller parts, or to remember how a similar problem was solved in the past, or to make a series of different attempts and then to compare and evaluate these—or to try to learn some completely different way to deal with such situations. This means that a low-level process inside your mind may engage so many higher-level ones that you end up in some new mental state that amounts to a different Way to Think.

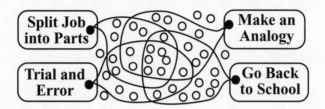

What if a person were to attempt to use several such Ways to Think at once? Then these would have to compete for resources, and that would need high-level management—which would usually choose one alternative. This could be one reason why it seems to us that our thoughts flow in serial, step-by-step streams—despite the fact that every such step must still be based on many smaller processes that operate simultaneously. In any case, this book will suggest that this so-called "stream of consciousness" is an illusion that comes because each higher-level part of one's mind has virtually no access to knowledge about what happens in most of one's other processes.

> Citizen: This idea of switching one's set of resources might explain the behavior of an insect or fish—but Charles doesn't switch, in the way you describe, to a totally different mental state. He just changes some aspects of how he behaves.

I completely agree. However, any theory has to begin with a highly simplified version of it—and even this trivial model could help to explain why human infants so frequently show such sudden changes in their states. But certainly, in later years, children develop more fluent techniques through which their resources can be aroused and suppressed to different extents—and this leads to more ability to combine both old instincts and new Ways to Think. Then, several of these can be active at once—and that's when we speak of our feelings as mixed.

## 1-6  Adult Emotions

"Behold the child, by nature's kindly law,
Pleas'd with a rattle, tickl'd with a straw:

Some livelier plaything gives his youth delight,
A little louder, but as empty quite:
Scarfs, garters, gold, amuse his riper stage,
And beads and pray'r books are the toys of age."
—*Alexander Pope, in* Essay on Man

When an infant gets upset, that change seems as quick as the flip of a switch.

A certain infant could not bear frustration, and would react to each setback by throwing a tantrum. He'd hold his breath and his back would contract so that he'd fall rearward on his head.

Yet several weeks later, that behavior had changed.

No longer completely controlled by his rage, he could also add ways to protect himself, so that when he felt a tantrum coming on, he'd run to collapse on some soft, padded place.

This suggests that, in the infant brain, only one *"Way to Think"* can work at a time, so that not many conflicts will arise. However, those infantile systems cannot resolve the conflicts we face in our later lives. This led our ancestors to evolve higher-level systems in which some instincts that formerly were distinct could now become increasingly mixed. But as we gained more abilities, we also gained new ways to make mistakes, so we also had to evolve new ways to control ourselves, as we'll see in Chapter 9-2.

We tend to regard a problem as "hard" when we've tried several methods without making progress. But it isn't enough just to know that you're stuck: you'll do better if you can recognize that you're facing some particular kind of obstacle. For if you can diagnose what Type of Problem you face, this can help you to select a more appropriate Way to Think. So this book will suggest that to deal with hard problems, our brains augmented their ancient Reaction-Machines with what we'll call *"Critic-Selector Machines."*

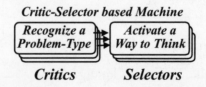

**Critic-Selector based Machine**

| Recognize a Problem-Type | Activate a Way to Think |
|---|---|
| **Critics** | **Selectors** |

The simplest versions of these would be the *"If→Do"* machines described in Section 1-4. There, when an *"If"* detects a certain real-world situation, its *"Do"* reacts with a certain real-world action. Of course, this means that simple *If→Do* machines are highly constrained and inflexible.

However, the "Critics" of Critic-Selector Machines will also detect situations or problems inside the mind such as serious conflicts between active resources. Similarly, the "Selectors" of Critic-Selector Machines don't just perform actions in the external world, they can react to *mental* obstacles by turning other resources on or off—thus switching to different Ways to Think.

For example, one such Way to Think would be to consider several alternative ways to proceed before selecting which action to take. Thus, an adult who encounters what might be a threat need not just react instinctively, but first could proceed to *deliberate* on whether to retreat or attack—by using high-level strategies to choose among possible ways to react. This way, one could make a thoughtful choice between becoming angry or becoming afraid. Thus when it seems appropriate to intimidate an adversary, one can make oneself angry deliberately—although one may not be aware that one is doing this.

How and where do we develop our higher-level Ways to Think? We know that during our childhood years, our brains go through multiple stages of growth. To make room for these, Chapter 5 will conjecture that this results in at least six levels of mental procedures, and this diagram will summarize our main ideas about how human minds are organized.

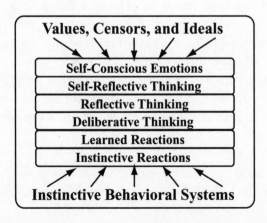

The lowest level of this diagram corresponds to the most common kinds of "instincts" with which our brains are equipped from birth. The highest levels support the sorts of ideas that we later acquire and call by names like *ethics* or *values*. In the middle are layers of methods we use to deal with all sorts of problems, conflicts, and goals; this includes much of our everyday commonsense thinking. For example, at the "deliberative" level, you might consider several different actions to take, then imagine the effects of each, and then compare those alternatives. Then, at the "reflective" levels, you might think about what you have done and wonder if the decisions you made were good—and finally, you might "self-reflect" about whether those actions were worthy of the ideals that you have set for yourself.

We all can observe the progression of our children's values and abilities. Yet none of us can recollect the early steps of our own mental growth! One reason for this could be that, during those times, we kept developing ways to build memories—and each time we switched to new versions of these, that made it difficult to retrieve (or to understand) the records we made in previous times. Perhaps those old memories still exist, but in forms that we no longer can comprehend—so we cannot remember how we progressed from using our infantile reaction-sets to using our more advanced Ways to Think. We've rebuilt our minds too many times to remember how our infancies felt!

## 1-7 **Emotion Cascades**

> Charles Darwin 1871: "Some habits are much more difficult to cure or change than others are. Hence a struggle may often be observed in animals between different instincts, or between an instinct and some habitual disposition; as when a dog rushes after a hare, is rebuked, pauses, hesitates, pursues again, or returns ashamed to his master; or as between the love of a female dog for her young puppies and for her master,—for she may be seen to slink away to them, as if half ashamed of not accompanying her master."

This chapter has raised some questions about how people could change their states so much. Let's look back to our first example of this: *When someone you know has fallen in love, it's almost as though a switch has been thrown, and a different program has started to run.* Our Critic-Selector model of mind suggests that such a change could result when a certain Selector activates a certain particular set of resources. Thus Charles's attraction to Celia becomes stronger because a certain Selector has suppressed most of his usual fault-finding Critics.

> Psychologist: Indeed, infatuations sometimes strike suddenly. But other emotions may slowly flow and ebb—and usually, in our later years, our mood shifts tend to become less abrupt. Thus, an adult may be slow to take offense, but may then go on to brood for months about even a small or imagined affront.

Our twenty-year-old tabby cat shows few signs of human maturity. At one moment she'll be affectionate, and seek out our companionship. But after a time, in the blink of an eye, she'll rise to her feet and walk away, without any sign of saying good-bye. In contrast, our twelve-year-old canine pet will rarely depart without looking back—as though he's expressing a certain regret. The cat's moods seem to show one at a time, but the dog's dispositions seem more mixed, and less as though controlled by a switch.

In either case, any large change in which resources are active will substantially alter one's mental state. Such a process might begin when one Selector resource directly arouses several others.

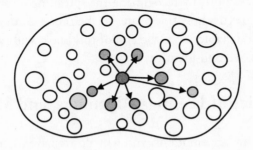

Then, some of those newly aroused resources may proceed to activating yet other ones—and if each such change leads to several more, this all could result in a large-scale "cascade."

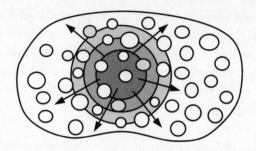

The further these activities spread, the more they will change your mental state, but, of course, this won't change everything. When Charles engages a new Way to Think, not all his resources will be replaced—so, in many respects he'll still be the same. He will still be able to see, hear, and speak—but now he'll perceive things in different ways, and may select different subjects to discuss. He may now have some different attitudes, but still will have access to most of his commonsense knowledge. He will still have some of the same plans and goals—but different ones will be pursued because they now have different priorities.

Yet despite all these changes, Charles will insist that he still has the same "identity." To what extent will he be aware of how his mental condition has altered? He sometimes won't notice those changes at all, but at other times, he may find himself asking questions like, *"Why am I getting so angry now?"* However, even to think of asking such questions, Charles's brain must be equipped with ways to "self-reflect" on some of his recent activities—for example, by recognizing the spreading of certain cascades. Chapter 4 will discuss how this relates to the processes that we call "consciousness," and Chapter 9, at the end of this book, will talk more about the concepts of Self and Identity.

## 1-8 Theories of Feelings, Meanings, and Machines

> Citizen: What are emotions, and why do we have them? What is the relation between one's emotions and one's intellect?

When we talk about a person's mind, we usually use the plural, *emotions,* but we always use the singular noun to speak about someone's *intellect.* However, this book will take the view that each person has multiple Ways to Think, and what we call "emotional" states are merely different examples of these. To be sure, we all grow up with the popular view that we have only a single Way to Think—called "logical" or "rational"—but that our thinking can be colored, or otherwise influenced by so-called emotional factors.

However, the concept of *Rational Thinking* is incomplete—because logic can only help us to draw conclusions from the assumptions that we happen to make—but logic, alone, says nothing about which assumptions we ought to make, so Chapter 7-4 will talk about more than a dozen other Ways to Think, in which logic plays only minor roles, while more of our mental power comes from finding useful analogies.

In any case, our Citizen's question illustrates our all-too-common tendency to try to divide any complex thing into two separate, complementary parts—such as *emotion* vs. *intellect.* However, Chapter 9-2 will argue that few such two-part distinctions really describe two genuinely different ideas. Instead, those "dumbbell" theories merely suggest a single idea *and then contrast it with everything else.* To avoid that, this book will take the view that, whenever you think about something complex, you should try to depict it with more than two parts, or else switch to some different Way to Think!

Citizen: Why would one want to think of oneself as though one were nothing more than a machine?

Saying that someone is like a machine has come to have two opposite meanings: (1) "to have no intentions, goals, or emotions," and (2) "to be relentlessly committed to a single purpose or policy." Each meaning suggests inhumanity, as well as a kind of stupidity, because excessive commitment results in rigidity, while lack of purpose leads to aimlessness. However, if the ideas in this book are right, both of those views will be obsolete, because we'll show ways to make machines that not only will have *persistence, aim,* and *resourcefulness,* but also will have hosts of checks and balances—as well as abilities to grow by further extending their abilities.

Citizen: But machines can't feel or imagine things. So even if we could make them think, would they not still be missing the sense of experience that gives meaning to our human lives?

We have many words that we can use to try to describe how we feel—but our culture has not encouraged us much to make theories of how those feelings work. We know that *anger* makes us more belligerent, and that *contented* people less often get into fights—but those emotions-words don't point to ideas about *how* those conditions affect our mental states.

We recognize this when we deal with machines: Suppose that one morning your car won't start, but when you ask your mechanic for help, you receive only this kind of reply: *"It appears that your car doesn't want to run. Perhaps it has become angry at you because you haven't been treating it well."* Clearly a "mentalistic" description like this won't help to explain how your car behaves. Yet we don't get annoyed when people use those kinds of words to describe events in our social lives.

However, if one wants to understand any complex thing—be it a brain or an automobile—one needs to develop good sets of ideas about the relationships among the parts inside. To know what might be wrong with that car, one must have enough knowledge to ask if there's something wrong with its starter switch, or whether the fuel tank has been completely drained, or whether some excessive strain has broken some shaft, or if some electrical circuit fault has completely discharged the battery. In the same way, one cannot get much from seeing a mind as a Single Self: one must study the parts to know the whole. So the rest of this book will argue that, for example, to understand why "being angry" feels the way it does, you will need much more detailed theories about the relationships among the parts of your mind.

> Citizen: If my mental resources keep changing so much, what gives me the sense that I'm still the same Self, no matter how happy or angry I get?

Why do all of us come to believe that somewhere, deep in the heart of each mind, there exists some permanent entity that experiences all our feelings and thoughts? Here is a very brief sketch of how I will try to answer this in Chapter 9:

> In our early stages of development, our low-level processes solve many small problems without any sense of how this happens. However, as we develop more levels of thought, those higher levels start to find ways to represent some aspects of our recent

thoughts. Eventually these develop into collections of "models" of ourselves.

A simple model of a person's Self might consist of just a few parts connected like those shown below. However, each person eventually builds more complex Self-models that represent ideas about, for example, one's social relationships, physical skills, and economic attitudes. So Chapter 9 will argue that when you say "Self," you are referring not to a single representation but to an extensive network of different models that represent different aspects of yourself.

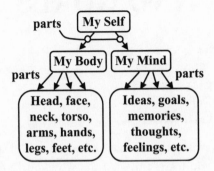

In the usual view of how human minds grow, each child begins with instinctive reactions, but then goes through stages of mental growth that give us additional layers and levels of processes. Those older instincts may still remain, but these new resources gain increasing control—until we can think about our own motives and goals and perhaps try to change or reformulate them.

But how could we learn which new goals to adopt? No infant could ever be wise enough to make good such choices by itself. So Chapter 2 will argue that our brains must come equipped with special kinds of machinery that help us, somehow, to absorb the goals and ideals of our parents and friends!

# 2
# ATTACHMENTS
# AND GOALS

## 2-1 Playing with Mud

"It's not just learning things that's important. It's learning what to
do with what you learn and learning why you learn things at all
that matters."
—*Norton Juster, in* The Phantom Tollbooth

A child named Carol is playing with mud. Equipped with a fork, a spoon,
and a cup, her goal is to bake a make-believe cake. Let's assume that at first
she is playing alone.

> *Playing alone.* Carol wants to fill her cup with mud, and first
> tries to do this with her fork, but this fails because the mud slips
> through. She feels frustrated and disappointed. But when she suc-
> ceeds by using her spoon, Carol feels satisfied and pleased.

What could Carol learn from this? She learns from her "trial and error"
experience that forks are not good for carrying mud. But then she learns
from success with a spoon that spoons are good tools for moving a fluid,
so she is likely to use this method the next time she wants to fill a cup.
Note that here Carol was working alone—and acquired new knowledge,

all by herself. *In the course of learning by trial and error, a person requires no teacher to help her.*

> A *stranger scolds*. Now a stranger appears and reproaches her: "That is a naughty thing to do." Carol feels anxious, alarmed, and afraid. Overcome by fear and the urge to escape, she interrupts her present goal—and runs to find her parent's protection.

What should Carol learn from this? This may have little or no effect on what she will learn about mud or about filling a cup—but she's likely to conclude that she has placed herself in an unsafe location. *Next time she'll play in some safer place.* Also, a sequence of scary encounters like this could make her become less adventurous.

> *Her mother scolds.* Carol returns to her mother for help, but instead of defense or encouragement, all she gets is a reproof: "What a disgusting mess you've made! See all the mud on your clothes and your face. I scarcely can bear to look at you!" Carol, ashamed, begins to cry.

What might Carol learn from this? She'll become less inclined toward playing with mud, whereas if her parent had chosen to praise her instead, she would have felt pride instead of shame—and in future times would be more inclined toward the same kind of play. *In the face of a parent's blame or reproach, she learns that her goal was not a good one to pursue.*

Think of how many emotional states our children engage in the thousand minutes of each waking day! In this very brief story we've touched upon *Satisfaction, Affection,* and *Pride*—feelings we think of as positive—and we also encountered *Shame, Fear, Disgust,* and *Anxiety*—conditions we think of as negative. What are the functions of all those mental conditions, and why do we so often classify them as positive and negative?

In most popular views of how learning works, the "positive" feelings that come with success are somehow involved with making us learn *new ways* to *behave*—whereas the "negative" feelings that failures bring make us learn *ways to not behave.* However, while this may apply to some animals, this idea of "learning by positive reinforcement" does not account for so much of how people learn because, frequently, *failures help more than successes do, when we try to acquire deeper ideas.*

We'll return to learning in Chapter 8, but this chapter will be more concerned with how we acquire new kinds of goals than with how we learn ways to achieve them. And because adult minds are so intricate, we'll begin by discussing what children do.

## 2-2 Attachments and Goals

Some of our strongest emotions come when we are near the persons to whom we've become attached. When we're praised or rejected by people we love, we don't just feel Pleasure or Dissatisfaction; instead, we tend to feel Pride or Shame. Of course, some functions of early attachments are clear: they help young animals to survive by providing nourishment, comfort, and defense. However, this section will argue that those particular feelings of Pride and Shame may play unique and peculiar roles in how humans develop new values and goals.

Most mammals, shortly after birth, can move and follow their mothers about—but humans are exceptions to that. Why did human infants evolve their much slower pace of development? Surely this was partly because their larger brains needed more time to mature. But also, as those more powerful brains led to more complex societies, our children no longer had time enough to learn from individual experience. Instead, we evolved ways to learn more efficiently by passing, directly from parent to child, enormous bodies of cultural knowledge. In short, we then became able to learn by "being told"! However, this did not become feasible until our big new brains evolved more powerful ways to represent knowledge—and then to "express" that knowledge in ways that eventually led to our languages.

To transmit that knowledge from parent to child, each party needed effective ways to engage and maintain each other's attention. Of course, our ancestors already had traits that helped to accomplish this; for instance, the infants of most animal species are born equipped with squeaks or squeals that arouse their parents from deepest sleep—and the brains of those parents contain machinery to force them to react to those cries. For example, those parents feel intense distress when they lose track of their infants' locations, while the infants have instincts that make them shriek whenever their parents become unavailable.

Also, as the length of their infancies grew, our children evolved increasing concern with how their parents reacted to them—and parents started to focus more on the growth of their children's values and goals. Thus, in the scene where Carol's mother reproached her, the child was likely to think thoughts like, *"I should not have wanted to play with mud, because that turned out to be an unsuitable goal."* In other words, Carol's shame caused her to change her goals instead of just learning ways to achieve them! Similarly, if her mother had praised her for her play, that praise could have led Carol to deepen her interest in material science and engineering.

It is one thing to learn *how to get* what you want—and another to learn *what you ought* to want. In our usual learning by trial and error, we improve our ways to achieve the goals that we already hold. However, when we "self-consciously" reflect on our goals (see Chapter 5-6), we're likely to change their priorities—and what I am suggesting is that self-conscious emotions like Pride and Shame play special roles; they help us learn *ends* instead of *means*. Thus, where trial and error teach us new ways to achieve the goals we already maintain—attachment-related blame and praise teach us which goals we should discard or retain. Listen to Michael Lewis depict some of the potent effects of shame:

> Michael Lewis 1995b: "Shame results when an individual judges his or her actions as a failure in regard to his or her standards, rules and goals and then makes a global attribution. The person experiencing shame wishes to hide, disappear or die. It is a highly negative and painful state that also disrupts ongoing behavior and causes confusion in thought and an inability to speak. The body of the shamed person seems to shrink, as if to disappear from the eye of the self or others. Because of the intensity of this emotional state, and the global attack on the self-system, all that individuals can do when presented with such a state is to attempt to rid themselves of it."

But when do people experience such intense and painful self-conscious sensations? Such feelings frequently come to us when we're in the presence of those we respect, or those by whom we wish to be respected; long ago this was recognized by another outstanding psychologist:

Aristotle b: "Now since shame is a mental picture of disgrace, in which we shrink from the disgrace itself and not from its consequences, and we only care what opinion is held of us because of the people who form that opinion, it follows that the people before whom we feel shame are those whose opinion of us matters to us. Such persons are: those who admire us, those whom we admire, those by whom we wish to be admired, those with whom we are competing, and whose opinion of us we respect."

This suggests that our values and goals are greatly influenced by the people to whom we become "attached"—at least in our earliest "formative" years. So the following sections will ask about how that type of learning might work, by discussing such questions as these:

What are the spans of those "formative" years?
To whom do our children become attached?
When and how do we outgrow attachments?
How do attachments help us establish our values?

You're almost always pursuing goals. Whenever you're hungry, you try to find food. When you sense danger, you strive to escape. When you've been wronged, you may wish for revenge. Sometimes your goal is to finish some work—or perhaps to seek ways to escape from it. We have a host of different words for such activities—such as to *try, wish, want, aim, strive,* and *seek*—but we rarely ask ourselves questions like these:

What are goals and how do they work?
What are the feelings that accompany them?
What makes some goals strong and others weak?
What could make an impulse "too strong to resist"?
What makes certain goals "active" now?
What determines how long they'll persist?

Here is one useful theory about when we use words like *want* and *goal:* You say that you *want* a certain thing *when you have an active mental process that works to reduce the difference between your present situation and one in which you possess that thing.* Here is a sketch of how a machine could do this:

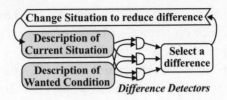

For example, every baby is born with two such systems for maintaining "normal" body temperature. One such "goal" is aroused when the child is too hot—and causes it to sweat, pant, stretch out, or vasodilate. However, when the baby is too cold, it will curl up, shiver, vasoconstrict, and/or raise its metabolic rate.

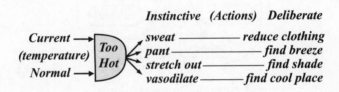

### Ways to React to Being Too Hot

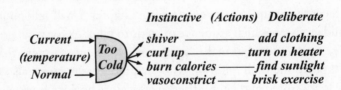

### Ways to React to Being Too Cold

Chapter 6-3 will show some more details of these kinds of goal-seeking machines.

When such processes work at low cognitive levels, at first we may not recognize them—for example, when you get too hot and start to sweat. However, when perspiration drips, you may notice this and deliberate: *"I must find some way to escape from this heat."* Then your higher-level knowledge suggests other actions that you could take—such as moving to a cooler place. Similarly, when you recognize that you feel cold, you might put on a sweater, turn on a stove, or begin to exercise (which can make your body produce ten times as much heat).

When you need to remove several differences, then this may require several steps. For example, suppose that you're hungry and *want* to eat, but you have only a can of soup. Then you'll *need* a tool to open that can, you'll *need* to find a bowl and a spoon, and you'll *need* a place where you can sit down to eat. So, each such *need* is a "subgoal" that comes from some difference between what you have now and what you desire.

A SIMPLE "SUBGOAL TREE"

Of course, to achieve several goals efficiently, you will need a plan, or else you might waste a good deal of time. It would be foolish to first sit down to eat before you have prepared your food, because then you would have to get up to start over again. Chapter 5 will talk about how one could envision which sequence of steps to take. As for what goals are, how they work, and what makes some goals seem more urgent than others, we'll postpone such questions till Chapter 6, where we'll also discuss how goals are stored and later retrieved, as well as how we learn new ways to achieve them. For now, we'll focus only on how we learn new goals and ideals.

## 2-3 Imprimers

"Never let your sense of morals prevent you from doing the
    right thing."
         —*Isaac Asimov*

While Carol was learning to fill her cup, she was annoyed when she failed with a fork, but she was pleased by success when using a spoon—so the next time she wants to fill a cup, she'll be more likely to know what to do. This is the most common conception of how people learn—that our reactions are "reinforced by success." This may seem commonsensical—but we need a theory of how that might work.

> Student: I suppose that her brain formed connections from her goal to the actions that helped her to achieve it.

Okay, but that is rather vague. Could you say more about how that might work?

> Student: Perhaps Carol starts with some goals just floating around—but when she succeeds by using her spoon, then she somehow connects "Fill Cup" to her "Use Spoon" goal. Also, when she fails with the fork, she makes a "do not" connection for "Use Fork," to keep from doing that again. So next time she wants to fill a cup, she'll first try the subgoal of using a spoon.

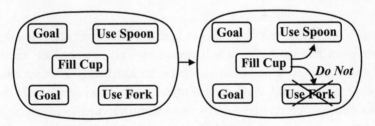

CONNECTING A SUBGOAL

That could be a good explanation of how Carol could connect a new subgoal to her original goal. And I approve of your mentioning *"do not"* connections because we must not only learn to do things that work, but we must also learn ways to avoid the most common mistakes. This suggests our mental connections should get "reinforced" by success, but should be suppressed whenever our actions don't work.

However, although this kind of "learning by trial and error" can connect new subgoals to existing goals, it does not explain how a person could learn completely new goals—or what we call *"values"* or *"ideals"*—that don't yet connect to existing ones. More generally, it does not cover the subject of how one could *learn what one "ought" to want.* I don't recall much discussion of this in academic psychology books, so here I'll conjecture that children do this in a special way that depends on how they interpret the reactions of *the persons to whom they are "attached."*

Our language uses a great many words for referring to our emotional states. When we described Carol's playing with mud, we used more than a dozen of them—*Affection, Alarm, Anxiety, Assurance, Disappointment, Disgrace, Disturbance, Frustration, Fear, Inclination, Pleasure, Pride, Satisfaction, Shame,* and *Sorrow.* This raises many questions about why we have such mental states at all—and why do we have so many of them? In particular, we need to ask: What makes Carol feel grateful and proud when her mother praises her? How might that "attachment bond" make her so much more concerned about her mother's regard for her? And how could this manage to "elevate" goals to make them seem more respectable?

> Student: My theory also fails to explain why praise from a stranger won't elevate goals. Why does this require the presence of—I can't think of the proper word for this—"a person to whom one is attached?"

I think it is remarkable that we do not have a special word for such a significant type of relationship! Psychologists cannot say "parent," or "mother," or "father" because a child can also become attached to another relative, nurse, or family friend. Psychologists often use the word *caregiver* for this— but as we'll see in Section 2-7, such attachments can form without physical care, so *caregiver* does not quite hit the target. So this book will introduce a new term derived from the old word *imprinting,* which has long been used by psychologists to refer to the processes that keep young animals close to their parents.

> *Imprimer:* An Imprimer is one of those persons to whom a child has become attached.

In most other species of animals, the function of infant attachment seems clear: remaining close to parents helps to keep their offspring safe. However, in humans it seems to have other effects; when Carol's Imprimer praises her, she feels a special thrill of pride that elevates her present goal to a status that is more "respectable." Thus, Carol's goal to work with mud may have begun as nothing more than a casual urge to play with materials in her environment. But—according to my conjecture here—her Imprimer's praise (or blame) appears to change the status of that goal into something more like an ethical value (or into one she regards as dishonorable).

Why might our brains use machinery that causes an Imprimer's praise to have an effect so different from that of praise that comes from a stranger? It is easy to see why this would have evolved: *if strangers could change your high-level goals, they could get you to do whatever they want, simply by changing what you, yourself, want to do!* Children with no defense against this would be less likely to survive, so evolution would tend to select those who have ways to resist that effect.

## 2-4 Attachment-Learning "Elevates" Goals

Michael Lewis 1995b: "Each of us has beliefs about what constitutes acceptable actions, thoughts and feelings. We acquire our standards, rules and goals through acculturation . . . and each of us has acquired a set appropriate to our particular circumstances. To become a member of any group, we are required to learn them. Living up to one's own internalized set of standards—or failing to live up to them—forms the basis of some very complex emotions."

When Carol's loved ones censure her, she feels that her goals are unworthy of her or that she is unworthy of her goals. And even in her later years, when her Imprimers are far from the scene, she still may wonder about how they might feel about her: *Would they approve of what I am doing? Would they praise the way I'm thinking now?* What kinds of machinery might we engage that makes us experience such concerns? Let's listen to Michael Lewis again:

Michael Lewis 1995b: "The so-called self-conscious emotions, such as guilt, pride, shame and hubris, require a fairly sophisticated level of intellectual development. To feel them, individuals must have a sense of self as well as a set of standards. They must also have notions of what constitutes success or failure, and the capacity to evaluate their own behavior."

Why would the growth of such personal values depend upon a child's attachments? Again we can see how this might have evolved: a child who lost her parents' esteem would be less likely to survive. Also, if those parents themselves want to earn the respect of their friends, they will want their children to "behave" in socially acceptable ways. So now we've seen several different ways in which our children might change themselves:

*Positive Experience:* When a method succeeds, learn to use that subgoal.
*Negative Experience:* When a method fails, learn to not use that subgoal.
*Aversion Learning:* When a stranger scolds, learn to avoid such situations.
*Attachment Praise:* When an imprimer praises, elevate your goal.
*Attachment Censure:* When an imprimer scolds, devalue your goal.
*Internal Imprimimg:* When an imprimer scolds, devalue your goal.

In Section 2-2 we saw a way to make a new goal depend upon an existing one, so that it could serve as a subgoal for it—the way we attached "Use Spoon" to "Fill Cup." But how could we "elevate" a goal *above* the ones we already hold? We can't leave it floating in empty space, because it would be useless for one to learn anything new unless one also connects it to ways to retrieve it when it is relevant. This means that we need some answers to questions about *to what each new goal should be attached, when and how it should be aroused,* and *how long to pursue it before giving up.* We'll also need more ideas about how a mind (or brain) could decide, when several goals are engaged at once, which of them should get higher priority. We'll talk about that in Chapter 5. And, of course, we'll need to clarify *what kind of thing a goal might be*—but we'll postpone that until Chapter 6.

However, here we'll start by focusing on how our goals might be organized. We already suggested in Chapter 1-6 that our mental resources might be located at various levels in what we described as an organizational layer cake.

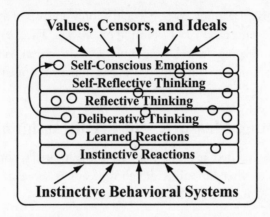

A SIX-LEVEL MODEL OF MENTAL ACTIVITIES

This six-level image is intended to be vague, because our brains are not so neatly arranged. However, this gives us a way to begin: imagine that the kinds of goals called "values" or "ideals" are attached to resources near the top, while our more infantile aims come from resources near the base of that cake. Then the arrow in this diagram suggests a possible meaning for "elevate."

> To "elevate" a goal could mean to copy, move, or link it to some higher location in that tower.

Then, our attachment-based learning scheme could be summarized in the form of this more general rule:

> *If* you detect Praise *and* an Imprimer is present, *Then* "elevate" your present goal.

But why should we need Imprimers at all—and why should we choose them so selectively, rather than simply elevate goals in response to anyone's censure or praise? Presumably, that rule evolved to include Imprimers because, as we noted before in Section 2-3, we all would be in danger if any stranger could reprogram our goals.

> Student: But surely that is not always true; I am not immune to compliments—even from persons I don't respect.

If attachment-based learning exists, it would be only one part of the story. Many other kinds of events can make us learn in other ways. The resourcefulness of the human mind comes from having multiple ways to deal with things—no matter that, from time to time, this causes bad things to happen to us.

## 2-5 Learning, Pleasure, and Credit Assignment

When Carol succeeded at filling her cup, she felt satisfaction and a sense of reward—but what functions did those feelings serve? It would seem that this process involved at least these three steps:

> Carol recognized that her goal was achieved.
> She felt some pleasure about her success.
> Then, somehow, that helped her to learn and remember.

Now we're happy that Carol felt gratified, but why can't she just "simply remember" which methods worked and which ones failed? What kinds of roles does pleasure play in establishing new memories?

The answer is that "remembering" is not simple at all. On the surface, it might seem easy enough—like dropping a note into a box and then taking it out when you need it. But when we look more closely, we see that this must involve many processes: you first must decide what items your note should contain, and find suitable ways to represent them—and then you must make some connections to them so that after you store those parts away, you'll be able to reassemble them.

> Student: Can't we explain all this with the old idea that, for each of our accomplishments, we just "reinforce" our successful reactions? In other words, we simply "associate" the problem we faced with the action or actions that solved it, by making one more *If→Then* rule.

That might help to describe what learning *does*—when seen from outside—but it doesn't explain how learning *works*. For neither *"the problem we faced"* nor *"the actions we took"* are simple objects that we can connect—so first your brain will need to construct descriptions for both that

*If* and that *Then*. Of course, the quality of what you learn will depend on the content of those two descriptions:

> The *If* must describe some relevant features and relationships of the situation you faced.
> The *Then* must describe some relevant aspects of the successful actions you took.

For Carol to learn effectively, her brain will need to identify which of her tactics turned out to help, and which of them only wasted her time. For example, after her struggle to fill the cup, should Carol attribute her final success to the shoes or the dress she was wearing then, or whether the weather was cloudy or clear, or to the location in which those events occurred? Let's suppose that she smiled while using that fork, but happened to frown when using that spoon; then what keeps her from learning irrelevant rules like, *"To fill a cup, it helps to frown"*?

In other words, when a person learns, it is not just a matter of "making connections" but is also a matter of making the structures that then get connected—which means that *we need to find some ways to represent not only those external events, but also the relevant mental events.* Thus, Carol will need some reflective resources to choose which of the Ways to Think that she used should be among the things that she remembers. No theory of learning can be complete unless it includes ideas about how we make these "credit assignments."

> Student: You still haven't explained where feelings come in, such as the pleasure that comes from Carol's success.

In everyday life, we routinely use terms like *Suffering, Pleasure, Enjoyment,* and *Grief*—but get stuck when we try to explain what these mean. The trouble comes, I think, because we think of such "feelings" as simple or basic, whereas each one involves intricate processes. For example, I suspect that what we call *"Pleasure"* is involved with the methods we use to identify *which of our recent activities should get credit for our recent successes.* Chapter 8-5 will talk about why human brains need powerful ways to make these kinds of "credit assignments," and Chapter 9-4 will argue that this may engage machinery that prevents us from thinking about other things. If so, we may have to recognize that many effects of *Pleasure* are negative!

## 2-6 Conscience, Values, and Self-Ideals

"I did not, however, commit suicide, because I wished to know
more of mathematics."
                —*Bertrand Russell*

One way people differ from animals (except, perhaps, the elephants) is in
the great lengths of our childhoods. This surely must be one reason why
no other species accumulates anything close to our human traditions and
values.

What kind of person would you like to be? Are you careful and cautious or brave and audacious? Do you follow the crowd, or prefer to lead?
Would you rather be tranquil or driven by passion? Such personal traits
depend, in part, upon each person's inheritance. But also they are partly
shaped by our networks of social attachments.

Once our human attachment bonds form, they begin to serve multiple
functions. First, they keep children close to their parents—and this provides
such services as nutrition, defense, and companionship. But also (if my theory is right) our attachments provide each child with new ways to rearrange
his priorities. Also, the self-conscious emotions that come with attachment
have other, very specific effects; Pride tends to make you more confident,
more optimistic, and more adventurous, while Shame makes you want to
change yourself so that you will never get into that state again.

What happens when a young child's Imprimers go absent? Shortly, we'll
see some evidence that this usually leads to severe distress. However, older
children better tolerate this, presumably because each child makes "internal
models" that help them to predict their Imprimers' reactions. Then each
such model would serve its child as an "internalized" system of values—and
this could be how people develop what we call *ethics, conscience,* or *moral
sense.* Perhaps Sigmund Freud had such a process in mind when he suggested that children can "introject" some of their parents' attitudes.

How might a child attempt to explain the praising and scolding that
now he will sense—even though no Imprimer is present? This might
make a child imagine that there was another person inside his mind—perhaps in the form of a made-up companion. Or perhaps the child might
embody that model into a certain external object, such as a rag doll or a
baby blanket. We know how distressed a child can get when deprived of
those irreplaceable objects.[1]

We should also ask what might happen if a child somehow gained more control over how that internal model behaves—so that now that *child could praise himself*—and thus select which goals to elevate—or else *that child could censure himself,* and thus impose new constraints on himself. This would make him "ethically autonomous" because he now can replace some of his imprinted value-sets. Then, if some of those older values persist in spite of attempts to alter them, this could lead to conflicts in which the child opposes his former Imprimers. However, if that child's brain were able to change all of its previous values and goals, then there would remain no constraints at all on what kind of person emerges from this—it could even be a sociopath.

What determines the kinds of ideals that develop inside each human mind? Every society, club, or group evolves some social and moral codes, by inventing various rules and taboos that help it decide what it ought to do or should not do. Those sets of constraints have awesome effects on every kind of organization; they shape the customs, traditions, and cultures of families, nations, professions, and faiths. They even can make those establishments value *themselves* above everything else—so that their members are happy to die for them, in endless successions of battles and wars.

How do people justify their ethical standards and principles? I'll parody several ideas about this.

> Social contractor: There is no absolute basis at all for the values and goals that people adopt. They merely are based on agreements and contracts that each individual makes with the rest of us.

> Sociobiologist: That "social contract" idea seems neat—except that no one remembers agreeing to it! Instead, I suspect that our ethics are mainly based on traits that evolved in our ancestors—just as in those breeds of dogs which were bred for becoming attached to their masters; in humans, we call this trait "loyalty."

Clearly, some of our traits are partly based on genes that we have inherited, but others spread in the form of contagious ideas that propagate from each brain to the next as parts of a cultural heritage.[2]

Theologian: There is only one basis for moral rules, and only my sect knows the way to those truths.

Optimist: I deeply believe that ethical values are self-evident. Everyone would be naturally good, except for being corrupted by being raised in abnormal environments.

Rationalist: I'm suspicious of terms like *deeply believe* and *self-evident* because they seem to mean only, *"I cannot explain why I believe this,"* and, *"I don't want to know how I came to believe it."*

To be sure, some thinkers might argue that we can use logical reasoning to deduce which high-level goals to choose. However, it seems to me that logic can only help us deduce what's implied by the assumptions we make—but it cannot help us to choose among which assumptions we ought to assume.

Mystic: Reasoning only clouds the mind by detaching it from reality. Until you learn not to think so much, you will never achieve enlightenment.

Psychoanalyst: Relying on "instincts" may only hide your unconscious goals and desires from you.

Existentialist: Whatever goal you happen to have, you should ask what purpose *that* purpose serves—and when you keep on doing this, soon you will see that your world is a total absurdity.

Sentimentalist: You're too concerned with goals and aims. Just watch some children and you will see curiosity and playfulness. They are not seeking any goals, but are enjoying the finding of novelties and the pleasures of making discoveries.

We like to think that a child's play is unconstrained, but when children appear to feel joyous and free, that may merely conceal their purposefulness; you can see this more clearly when you attempt to drag them away from their chosen tasks. In fact, the "playfulness" of childhood is the most demanding teacher that one could have; it makes us explore our world to see what's there, to try to explain what all those structures are, and to imagine what else could possibly be. Exploring, explaining, and learning must be among a child's most obstinate drives—and never again in those children's lives will anything push them to work so hard.

## 2-7 Attachments of Infants and Animals

*"We want to make a machine that will be proud of us."*
*—Danny Hillis*

The young child Carol loves to explore, but she also likes to stay near to her mother—so, if she discovers that she is alone, she'll soon cry out and look for her mom. Also, whenever the distance between them grows, she quickly moves herself closer. And whenever there's cause for fear or alarm—such as when a stranger approaches her—that same behavior will appear, even when her mother is near.

Presumably, this dependency stems from our infantile helplessness: no human infant would long survive if she could escape from parental care, but that rarely happens because our infants can hardly move themselves at all. Fortunately, not much harm results from that because of an opposite bond that we also evolved: Carol's mother is almost always aware of what is happening to her daughter—and her full attention will be engaged at the slightest suspicion that something is wrong.

Clearly, each infant's survival depends on becoming attached to persons concerned with his welfare. So in older times it was often assumed that *children would attach themselves to the persons who gave them physical care,* and this is why most psychologists called such a person a "caregiver"—instead of using some word like *Imprimer.* However, physical care may not be the most critical factor, as suggested by John Bowlby, who pioneered systematic research on infant attachment.

> John Bowlby 1973a: "That an infant can become attached to others of the same age, or only a little older, makes it plain that attachment behavior can develop and be directed towards [persons who have] done nothing to meet the infant's physiological needs."[3]

Then what are the functions of our children's attachments? Bowlby's main concern was to refute the then popular view that attachment's primary function was to ensure a dependable source of food. Instead, he argued that nutrition played a smaller role than did physical security, and that (in our animal ancestors) our attachments served mainly to ward off attacks from predators. Here is a paraphrase of his argument:

> First, an isolated animal is much more likely to be attacked than is one that stays bunched together with others of its kind. Second, attachment behavior is especially easy to arouse in animals that— by reason of age, size, or conditions—are especially vulnerable to predators. Third, this behavior is strongly elicited in situations of alarm, which are commonly ones in which a predator is sensed or suspected. No other theory fits these facts.

I suspect that this was largely correct for most animals, but does not sufficiently emphasize how human attachments also help us to acquire our high-level values and goals. This still leaves us with the question of what are the factors that determine to whom our children will become attached? Physical nurture can play a significant role (by providing occasions for children to become attached)—but Bowlby concluded that, usually, these two other factors were more important:

> The quickness with which the person responds, and
> The intensity of that interaction.

In any case, a child's Imprimers will usually include his parents, but could also include his companions and friends. This suggests that parents should take special care to examine their offspring's acquaintances—and, especially, the ones who are most attentive to them. (For example, when selecting a school, a parent ought to scrutinize not only the staff and curriculum, but also the goals that its pupils pursue.)

What happens when a child is deprived of Imprimers? Bowlby concluded that this eventually leads to a special variety of fear, and a powerful impulse to find that Imprimer.

> John Bowlby 1973b: "Whenever a young child . . . is separated from his mother unwillingly he shows distress; and should he also be placed in a strange environment and cared for by a succession of strange people such distress is likely to be intense. The way he

behaves follows a typical sequence. At first he protests vigorously and tries by all the means available to him to recover his mother. Later he seems to despair of recovering her but nonetheless remains preoccupied with her and vigilant for her return. Later still he seems to lose his interest in his mother and to become emotionally detached from her."

Bowlby goes on to describe what happens when the mother comes back:

John Bowlby 1973b: "Nevertheless, provided the period of separation is not too prolonged, a child does not remain detached indefinitely. Sooner or later after being reunited with his mother his attachment to her emerges afresh. Thenceforward, for days or weeks, and sometimes for much longer, he insists on staying close to her. Furthermore, whenever he suspects he will lose her again he exhibits acute anxiety. . . .

"The very detailed observations made by Jane Goodall of chimpanzees in the Gombe Stream Reserve in central Africa show not only that anxious and distressed behavior on being separated, as reported of animals in captivity, occurs also in the wild but that distress at separation continues throughout chimpanzee childhood. . . . Not until young are four and a half years of age are any of them seen traveling not in the company of mother, and then only rarely."

Also, it was discovered that when young children are deprived of Imprimers for more than a few days, they often show signs of impairment for much longer times.

John Bowlby 1973b: "From all these findings we can conclude with confidence not only that a single separation of no longer than six days at six months of age has perceptible effects two years later on rhesus infants, but that the effects of a separation are proportionate to its length. A thirteen-day separation is worse than a six-day; two six-day separations are worse than a single six-day separation."[4]

To some, it may seem surprising that even badly mistreated children (and monkeys) may remain attached to abusive Imprimers (Seay 1964). Perhaps this might not seem so strange in the light of Bowlby's claim that attachment depends on *"the quickness with which the person responds, and the intensity of that interaction"*—because abusive persons also often excel in exactly those characteristics!

We see similar behaviors in our various primate relatives—such as orangutans, gorillas, and chimpanzees—as well as in our more distant cousins, the monkeys. We should also note Harry Harlow's discovery that, given no other alternative, a monkey will become attached to an object that has no behavior at all, but still has some "comforting" characteristics. This would seem to confirm Bowlby's view that attachment does not stem from physiological needs—unless we amend this to include what Harlow calls *"comfort contact."* (See Harlow 1958.)

When the mother and child have more distance between them, they maintain their connection with a special "hoo" whimper to which the other promptly responds—as Jane Goodall (1968) herself reports:

> "When the infant [chimpanzee] . . . begins to move from its mother, it invariably utters this sound if it gets into any difficulty and cannot quickly return to her. Until the infant's locomotion patterns are fairly well developed the mother normally responds by going to fetch it at once. The same sound is used by the mother when she reaches to remove her infant from some potentially dangerous situation or even, on occasion, as she gestures it to cling on when she is ready to go. The 'hoo' whimper therefore serves as a fairly specific signal in reestablishing mother-infant contact."

What happens in other animals? Early in the 1930s, Konrad Lorenz, a great observer of animals, found that a recently hatched chicken, duck, or goose will become "attached" to the first large moving object it sees, and will subsequently follow that object around. He called this "imprinting" because it occurs with such remarkable speed and permanence. Here are some of his observations:

> Imprinting begins soon after hatching.
> The chick quickly starts to follow the moving object.
> The period for imprinting ends a few hours later.
> The effect of imprinting is permanent.

To what kinds of objects do chicks get attached? Those moving objects are usually parents, but if the parents have been removed, then the object could be a cardboard box or a red balloon—or could even be Lorenz himself. Then, during the next two days, as the gosling follows its parents, it somehow learns to recognize them as individuals and not follow any other geese. Now when it loses contact with the mother, it will cease to feed or examine things, and instead will search and make piping sounds (like the "hoo" signals in Jane Goodall's notes) as though distressed at being lost. Then the parent responds with a special sound—and Lorenz observes that this response must come quickly to establish imprinting. (Later this call is no longer required, but in the meantime it serves to protect the chick against becoming attached to an unsuitable object, such as the moving branch of a tree.) In any case, these types of birds can feed themselves soon after they hatch, so imprinting is independent of being fed.

To what extent did human attachment-based learning evolve from older prehuman forms of imprinting? Humans, of course, are different from birds, yet the infants of both share some similar needs—and there may have been much earlier precursors of this; for example, Jack Horner (1998) has discovered that some dinosaurs constructed clusters of bird nest–like structures.

Returning to the human realm, we should ask how infants distinguish potential Imprimers. Although some researchers have reported that infants can recognize the mother's voice even before the time of birth, it is generally thought that newborns first learn mainly through touch, taste, and smell—and later distinguish the sound of a voice and react to the sight a face. One might assume that the latter depends on discerning such features as eyes, nose, and mouth, but there seems to be more to the story than that:

Francesca Acerra 1999: "4-day-old neonates look longer at their mother's face than at a stranger's face—but not when the mother wears a scarf that hides the hair contour and the outer contour of the head."

This suggests that those infants may react less to the features of the face, and more to its larger-scale, overall shape; it was not until two or three more months that Acerra's subjects were able to distinguish particular faces.[5] This suggests that our visual systems may use different sets of processes at different stages of development—and perhaps the ones that operate first serve mainly to attach the mother to her child! In any case, Konrad Lorenz was amazed by what his goslings *failed* to distinguish:

> Konrad Lorenz 1970: "The human imprinted gosling will unequivocally refuse to follow a goose instead of a human, but it will not differentiate between a petite, slender young girl and a big old man with a beard. . . . It is astounding that a bird reared by, and imprinted to, a human being should direct its behavior patterns not towards one human but towards the species Homo sapiens."

(I do not find this to be so strange because all geese look so much the same to me.) Perhaps more significant is Lorenz's claim that adult sexual preference may be established at this early time, though it appears only much later in behavior.

> "A jackdaw for which the human has replaced the parental companion, will thus direct its awakening sexual instincts not specifically towards its former parental companion, but . . . towards any one relatively unfamiliar human being. The sex is unimportant, but the object will quite definitely be human. It would seem that the former parental companion is simply not considered as a possible mate."

Could such delays be relevant to human sexual preferences? Studies have shown that, after more contact, some of those birds will eventually mate with other members of their species. However, this still is a serious obstacle to repopulating endangered species, so now it is standard policy to minimize human contact with new chicks before they are released.

All of this could help to explain why we evolved our extended infantile helplessness: children who too soon went off by themselves could not become wise enough to survive—and so, we had to extend the time during which those children were forced to learn from Imprimers.

## 2-8  Who Are Our Imprimers?

A Jackdaw, seeing Doves in a place with much food, painted himself white to join them. The Doves, as long as he did not speak, assumed that he was another Dove and admitted him to their cote. But when one day he forgot not to speak, they expelled him because his voice was wrong—and when he returned to his Jackdaw tribe they expelled him because his color was wrong. So desiring two ends, he obtained neither.

—*Aesop's Fables*

When do attachments begin and end? Even young infants soon start to behave in distinctive ways when in their mothers' presence. However, it is usually not till near the first year's end that the child protests against separation, and begins to learn to become disturbed at a sign that his Imprimer *intends* to depart—e.g., reaching for an overcoat. This is also the time when most children begin to show fears of unusual things. Both this fear of strange things and that fear of separation begin to decline in the child's third year—so that now the child can be sent to school. However, we do not see the same decline in the roles of those other, self-conscious, attachment-based feelings. These persist for longer times and sometimes, perhaps, for the rest of our lives.

John Bowlby 1973a: "During adolescence . . . other adults may come to assume an importance equal to or greater than that of the parents, and sexual attraction to age-mates begins to extend the picture. As a result individual variation, already great, becomes even greater. At one extreme are adolescents who cut themselves off from the parents; at the other are those who remain intensely attached and are unwilling or unable to direct their attachment behavior to others. Between the extremes lie the great majority of adolescents whose attachments to parents remain strong but whose ties to others are of much importance also. For most individuals the bond to parents continues into adult life and affects behavior in countless ways. Finally in old age, when attachment behavior can no longer be directed to members of an older generation, or even the same generation, it may come instead to be directed towards members of a younger one."

What happens in other animals? In those that do not remain in herds, attachment frequently persists only until the offspring can live by themselves. In many species it's different for females; in many species the mother will actively drive the young ones away as soon as a new litter is born (perhaps because of evolutionary selection against inbreeding)—while in other cases attachment will stay until the time of puberty or even later for females. Bowlby mentions a phenomenon that results from this:

> "In the female of ungulate species (sheep, deer, oxen, etc.), attachment to mother may continue until old age. As a result a flock of sheep, or a herd of deer, is built up of young following mother following grandmother following great grandmother and so on. Young males of these species, by contrast, break away from mother when they reach adolescence. Thenceforward they become attached to older males and remain with them all their lives except during the few weeks of each year of the rutting season."

Of course, other species evolve different strategies that are suited for different environments; for example, the size of the flock may depend on the character and prevalence of predators, etc.

When does that imprinting period end? R. A. Hinde discovered that chicks like the ones that Lorenz observed eventually become fearful of unfamiliar moving things. This led Hinde to suggest that time for imprinting comes to a stop only when this new kind of fear forestalls any further "following." Similarly, many human babies show a long period of fear of strangers that begins near the start of the second year.[6]

## 2-9 Self-Models and Self-Discipline

To solve a hard problem, you must work out a plan—but then you need to carry it out; it won't help to have a multistep plan if you tend to quit before it is done. This means that you'll need some "self-discipline"—which in turn needs enough self-consistency that you can predict, to some extent, what you're likely to do in the future. We all know people who make clever plans but rarely manage to carry them out because their models of what they will actually do don't conform enough to reality. But how could a trillion-synapse machine ever become predictable? How did our brains come to manage themselves in the face of their own great complexity? The answer must be that we learn to represent things in extremely compact, yet useful ways.

Thus, consider how remarkable it is that we can describe a person with words. What makes us able to compress an entire personality into a short phrase like "Joan is tidy," or "Carol is smart," or "Charles tries to be dignified"? Why should one person be *generally* neat, rather than be tidy in some ways and messy in others? Why should traits like these exist? In Chapter 9-2: Personality Traits, we'll see some ways in which such things could come about:

> In the course of each person's development, we tend to evolve certain policies that seem so consistent that we (and our friends) can recognize them as features or traits—and we use these to build our self-images. Then when we try to formulate plans, we can use those traits to predict what we'll do (and to thus discard plans that we won't pursue). Whenever this works, we're gratified, and this leads us to further train ourselves to behave in accord with these simplified descriptions. Thus, over time our imagined traits proceed to make themselves more real.

Of course, these self-images are highly simplified; we never come to know very much about our own mental processes, and what we call traits are only the seeming consistencies that we learn to use for describing ourselves. (See Chapter 9-2.) However, even these may be enough to help us conform to our expectations so that this process can eventually provide us with useful models of our own abilities.

We all know the value of having friends who usually do what they say they will do. But it's even more useful to be able to trust *yourself* to do what you've asked yourself to do! And perhaps the simplest way to do that is to make yourself consistent with the caricatures that you've made of yourself—by behaving in accord with self-images described in terms of sets of traits.

But how do those traits originate? Surely these can be partly genetic; we can sometimes perceive newborn infants to be more placid or more excitable. And, of course, some traits could be the chance results of developmental accidents. However, other traits seem more clearly acquired from contacts with one's Imprimers.

Is there some risk in becoming attached to too many different personalities? If a child has only a single Imprimer—or several that share very similar values—it won't be too hard for that child to learn which behaviors will usually be approved. But what is likely to happen when a child acquires several Imprimers who have conflicting sets of ideals? That could lead to the child's attempting to model herself on several different sets of traits—which could impair her development, because a person with coherent goals should usually do better than one encumbered by conflicting ones. Also, if you behave consistently, then, as we'll suggest in Chapter 9-2, this can help make other persons feel that they can depend on you. Nevertheless, Chapter 9 will argue that we should not expect a person to form only a single, coherent self-image: in fact, we each construct multiple models of ourselves, and learn when it's useful to switch among them.

In any case, if you changed your ideals too recklessly, you could never predict what we might want next: you would never be able to get much done if you could not "depend on yourself." However, on the other side, one needs to be able to compromise; it would be rash to commit to some long-range plan with no way to later back out of it. And it would be especially dangerous to change oneself in ways that prevent one from ever changing again. So it would seem that human beings find different ways to deal with this: some children end up with too many constraints, while other children adopt more ambitions than they will ever have time to implement.

Also, our Imprimers may feel the need to prevent their devotees from attaching themselves to persons of "dubious character." Here is an instance in which a researcher had to become concerned with who might influence his machine!

> In the 1950s, Arthur Samuel, a computer designer at IBM, developed a program that learned to play checkers well enough to defeat several excellent human players. Its quality of play improved when it competed with its superiors. However, games against inferior players tended to make its performance get worse—so much that its programmer had to turn its learning off. In the end, Samuel allowed his machine to play only against transcripts of master-class championship games.

We sometimes see this carried to extremes; consider how zealots recruit for their cults: they remove you from all your familiar locations and persuade you to break all your social attachments—including all your family ties. Then once you've been detached from your friends it becomes easy to sabotage all your defenses—and then you are ready to be imprimed by their local prophet, seer, or diviner, who has mastered some ways to implant new ideals into your anxious and insecure mind.

We face the same prospect in other realms. While your parents are concerned for your welfare, businesspersons may have more interest in promoting the wealth of their firms. Religious leaders may wish you well, yet be more concerned for their temples and sects. And when leaders appeal to your national pride, they may also expect you to lay down your life to defend some ancient boundary line. Each organization has its own intentions, and uses its members to further them.

> Individualist: I hope you don't mean that literally. An organization is nothing more than the circle of persons involved with it. It cannot have any goals of its own, but only those that its members hold.

What could it mean when someone suggests that some system has an intention or goal? Chapter 6-3 will discuss some conditions in which a process could seem to have motives.

## 2-10 Public Imprimers

We've discussed how attachment-based learning might work when a child is close to an Imprimer—but this might also relate to what happens when someone "catches the public's eye" by appearing in broadcast media. A straightforward way to promote a product would be to present good evidence for its virtue or value. However, we often see "testimonials" that only claim that a certain "celebrity" person approves of it. Why would this method work so well to influence someone's personal goals?

Perhaps part of the answer can be found by asking what factors might make those "celebrities" so popular. Attractive physical features may help, but also, most actors and singers have special skills: they are experts at feigning emotional states. Competitive athletes, too, are proficient deceivers—as well as are most popular leaders. But perhaps the most effective technique could be based on knowing ways to make each listener feel a sense that "this important person is speaking to *me*." This would make listeners feel more involved and, accordingly, more compelled to respond—no matter that they are only hearing a monologue!

Not everyone can control a mob. What techniques could one use to engage a very broad range of different minds? The popular term *charisma* has been defined to be *"a rare personal quality attributed to leaders who arouse popular devotion or enthusiasm."* When popular leaders mold our goals, could they be exploiting some special techniques through which they can establish rapid attachments?

> Politician: It usually helps for the speaker to have large stature, deep voice, and confident manner. However, although great height and bulk attract attention, some leaders have been diminutive. And while some powerful orators intone their words with deliberate measure, some leaders and preachers rant and shriek and still manage to grip our attention.

> Psychologist: Yes, but I see a problem with this. Earlier you mentioned that "speed and intensity of response" were important for making attachments. But when someone makes a public pronouncement, there is no room for those critical factors because the speaker cannot respond individually to each listener.

Rhetoric can create that illusion. A well-paced speech can seem "interactive" by raising questions in listeners' minds—*and then answering them at just the right time.* You can do this by interacting, inside your mind, with some "simulated listeners"—so that at least some of your audience will feel that they got an attentive response, although there was no genuine dialogue. Another trick would be to pause just long enough to make listeners to feel that they ought to react—but not to give them quite enough time to think of objections to your messages. Finally, an orator does not need to control everyone in the audience—because if you can recruit enough of them, then "peer pressure" may bring in the rest of them.

Conversely, a crowd could take over control of a more sensitive and responsive person in charge. Listen to one great performer who tried to avoid the influence of his audience:

> Glenn Gould: "For me, the lack of an audience—the total anonymity of the studio—provides the greatest incentive to satisfy my own demands upon myself without consideration for, or qualification by, the intellectual appetite, or lack of it, on the part of the audience. My own view is, paradoxically, that by pursuing the most narcissistic relation to artistic satisfaction one can best fulfill the fundamental obligation of the artist of giving pleasure to others."[7]

Finally, we also should note that a child could even become attached to an entity that doesn't exist—such as a person in some legend or myth, a fictional character in a book, or an imaginary animal. A person can even become attached to an abstract doctrine, dogma, or creed—or to an icon or image that represents it. Then those imagined entities could serve as "virtual mentors" inside their worshippers' minds. After all, when you come right down to it, *all* our attachments are made to fictions; you never connect to an actual person, but only to the models you've made to represent your conceptions of them.

So far as I know, this theory of how impriming works is new, although Freud must have imagined some similar schemes. What kind of experiments could show whether on not our brains use processes like this? New instruments that show events in brains might help, but experiments on human attachments might be deemed to be unethical. However, today we have an alternative: to write computer programs to simulate this. Then, if those programs behave in humanlike ways, this would show that our theory is plausible. But then, the computers might complain that we have not been treating *them* properly.

This chapter addressed some questions about how people choose which goals to pursue. Some of our goals are instinctive drives that come with our genetic inheritance, while others are subgoals that we learn (by trial and error) to accomplish goals that we already hold. As for our higher-level goals, this chapter conjectured those are produced by special machinery that makes us adopt the values of the parents, friends, or acquaintances to whom we become "attached," because they respond actively to our needs—and thereby induce in us such "self-conscious" feelings as Shame and Pride.

At first, those "Imprimers" must be near to us, but once we make "mental models" of them, we can use those models to "elevate" goals even when those Imprimers are absent—and eventually, these models become what we call *conscience, ideals,* or *moral codes.* Thus, attachments teach us ends, not means—and thus impose our parents' dreams on us.

We'll come back to this notion near the end of this book, but next we'll look more closely at the clusters of feelings that we know by names like *Hurting, Grief,* and *Suffering.*

# 3

# FROM PAIN TO
# SUFFERING

## 3-1 Being in Pain

> Charles Darwin 1872: "Great pain urges all animals, and has
> urged them during endless generations, to make the most violent
> and diversified efforts to escape from the cause of suffering. Even
> when a limb or other separate part of the body is hurt, we often
> see a tendency to shake it, as if to shake off the cause, though this
> may obviously be impossible."

What happens when you stub your toe? You've scarcely felt the impact yet,
but you catch your breath and start to sweat—because you know what's
coming next: a dreadful ache will tear at your gut; and all other goals will
be brushed away, replaced by your wish to escape from that pain.

How could such a simple event distort all your other thoughts so
much? What could make the sensation called *"Pain"* lead one into the
state we call *"Suffering"*? This chapter proposes a theory for this: any pain
will activate the goal *"Get rid of that pain"*—and achieving this will also
make that goal go away. However, if that pain is intense and persistent
enough, this will arouse yet other resources that tend to suppress your
other goals—and if this grows into a large-scale "cascade," there won't be
much left of the rest of your mind.

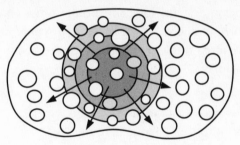

A SPREADING CASCADE

Of course, sometimes a pain is just a pain; if it doesn't last long or it's not too intense, then it won't escalate into suffering. Besides, you can usually muzzle a pain for a time, by trying to think about something else. Sometimes you can even make it hurt less by thinking about the pain itself: just focus your attention on it, evaluate its intensity, and try to regard its qualities as interesting novelties. But this provides only a brief reprieve because, whatever diversions you try, pain continues to gripe and complain, like a nagging, frustrated child; you can think about something else for a time, but will soon again be distracted to its demands.

> Daniel Dennett 1978: "If you can make yourself study your pains (even quite intense pains) you will find, as it were, no room left to mind them: (they stop hurting). However, studying a pain (e.g., a headache) gets boring pretty fast, and as soon as you stop studying them, they come back and hurt, which, oddly enough, is sometimes less boring than being bored by them and so, to some degree, preferable."

In any case, we should be thankful that pain evolved, because it protects our bodies from harm, first by making one try to remove its cause, and then by helping the injured part to rest and repair itself by keeping one from moving it. Here are some other ways in which pain protects us from injury.

> Pain makes you focus on the body parts involved.
> It makes it hard to think about anything else.
> Pain makes you move away from its cause.
> It makes you want that state to end, while teaching you,
>      for future times, not to repeat the same mistake.

Yet instead of being grateful for pain, people often complain about it. *"Why are we cursed,"* pain's victims ask, *"with such unpleasant experiences?"* And although we often think of pain and pleasure as opposites, they have many similar qualities:

> Pleasure often makes you focus on the body parts involved.
> It makes it hard to think about anything else.
> It makes you draw closer to its cause.
> It makes you want to maintain that state, while teaching you,
>     for future times, to keep repeating the same "mistake."

All this suggests that both pleasure and pain engage some of the same kinds of machinery; both constrict one's range of attention, both have connections with how we learn, and both reduce the priorities of almost all one's other goals. In view of these similarities, an alien from outer space might wonder why people like pleasure so much—yet display so little desire for pain.

> Alien: Why do you humans complain about pain?
> Human: We don't like pain because it hurts.
> Alien: Then explain to me what "hurting" is.
> Human: Hurting is simply the way pain feels bad.
> Alien: Then please tell me what you mean by "feels bad."

At this point the human might insist that feelings are so basic and elemental that there simply is no way to explain them to someone who has not experienced them.

> Dualist philosopher: Science can explain a thing only in terms of other, yet simpler things. But subjective feelings like pleasure or pain cannot be reduced to smaller parts.

However, in Chapter 9 I'll argue that feelings are not basic at all, but are processes made of many parts—and that once we recognize their complexity, this will help us find ways to explain what feelings are and how they work.

## 3-2 How Does Pain Lead to Suffering?

We often speak of Hurting, Pain, and Suffering as though they were more or less the same, and differ mainly in degree. However, while the effects of transient discomforts are brief, the longer that Pain remains intense, the longer those cascades will continue to grow, and your efforts to think will deteriorate—so that goals that seemed easy in normal times get increasingly harder to achieve, as more resources become disturbed or suppressed. Then we use words like *Suffering, Anguish,* and *Torment* to describe what happens when persistent pain comes to disrupt so many other parts of your mind that you can barely think about anything but how this condition is impairing you.

> "I'm so something that I can't remember what it's called."
> —*Miles Steele (age 5)*

In other words, it seems to me that a major component of *Suffering* is the frustration that comes with the loss of your options; it is as though most of your mind has been stolen from you, and your awareness of this only makes things seem worse. For example, I have heard *Suffering* likened to a balloon that keeps dilating inside one's mind until there's no more room for its usual thoughts. This image suggests, among other things, that one has lost so much "freedom of choice" that one has become a prisoner. Here are a few of the sorrows that come when *Suffering* imprisons us:

Anguish of losing mobility
Resentment of not being able to think
Dread of becoming disabled and helpless
Shame of becoming a burden to friends
Remorse at dishonoring obligations
Dismay at the prospect of failure
Mortification of seeming abnormal
Horror and fear of impending death

Of course, we also lose some "freedom of choice" when we get into any particular mental state, because then we're constrained by the goals that accompany it. We never have enough time to do all the things we want to do—and every new idea or ambition is sure to conflict with some previous ones. Most times, we don't mind those conflicts much, because we feel that we're still in control—partly because we usually know that if we do not like the result, we still can go back and try something else.

However, whenever an aching Pain breaks in, all our projects and plans get thrust aside, as though by an external force—and then all we have left are desperate schemes for finding ways to escape from the Pain. Pain's imperatives can serve us well when they help us to deal with emergencies—but when a pain cannot be relieved, it can turn into a catastrophe.

> The primary function of Pain is to compel one to remove what is causing it—but in doing so, it tends to disrupt most of a person's other goals. Then, if this results in a large-scale cascade, we use words like Anguish or Suffering to describe what remains of its victim's mind.

Indeed, Suffering can affect you so much that your friends may see you as being replaced by a different personality. It may even make you cry out and beg for help, as though you've regressed to becoming an infant again. Of course, you may still seem the same to yourself, because you have the sense of still having access to the same memories and abilities—although they no longer seem of much use to you.

> "Life is full of misery, loneliness, and suffering—and it's all over much too soon."
> —*Woody Allen*

## 3-3  The Machinery of Suffering

> "The restless, busy nature of the world, this, I declare, is at the root of pain. Attain that composure of mind, which is resting in the peace of immortality. Self is but a heap of composite qualities, and its world is empty like a fantasy."
> —*Buddha*

Here is an example of what can happen when a person becomes a victim of pain:

> Yesterday Joan picked up a heavy box, and today there's a terrible pain in her knee. She's been working on an important report,

which she has to present at a meeting tomorrow. *"But if this keeps up,"* she hears herself think, *"I won't be able to take that trip."* She decides to visit her medicine shelf to get a pill that might bring some relief—but a stab of pain stops her from getting up. Joan clutches her knee, catches her breath, and tries to think about what to do next—but the pain so overwhelms her that she cannot focus on anything else.

*"Get rid of me,"* Joan's pain insists—but how does Joan know that it comes from her knee? Each person is born equipped with nerves that connect from each part of the skin to several different "maps" in the brain, such as this one in the sensory cortex, as depicted here.[1]

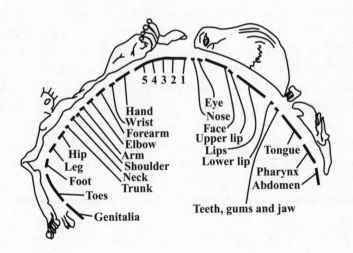

However, we are not born with similar ways to represent signals that come from our internal organs, and this could be why we find it hard to describe those pains that are not located near our skin; presumably, no such maps evolved because we would have had little use for them. For before the advent of modern surgery, we had no way to repair or protect a damaged liver or pancreas except by guarding one's entire belly, so all one needed to know was that one was having a bellyache. Similarly, we had no remedies that applied to specific places inside our brains, so it would not have helped to recognize that a pain came from one's cortex or thalamus.

As for the sense of Pain itself, our scientists know quite a lot about the first few events that result when a part of the body is traumatized: here is a typical attempt to describe what happens after that:

> Pain begins when special nerves react to pressure, cold, heat, etc., or to chemicals released by injured cells. Then the signals from those nerves rise up through the spinal cord to the thalamus, which relays them to other parts of your brain—in ways that seem to involve hormones, endorphins, and neurotransmitters. Eventually, some of those signals reach your limbic system, and this results in emotions like Sadness, Anger, and Frustration.

However, to understand how Pain can then lead to changes in our mental states, it doesn't much help to know only where various functions take place in the brain; we would also need to know what each one of those regions of the brain *does*—and how its processes interact with the other parts that are connected to it. Are any particular parts of the brain in charge of our hurting and suffering? Apparently so, to some extent, as cautiously noted by Ronald Melzack and Patrick Wall, who pioneered theories of how Pain works:

> Melzack and Wall 1965: "An area within the functionally complex anterior cingulate cortex has a highly selective role in pain processing, consistent with an involvement in the characteristic emotional/ motivational component (unpleasantness and urgency) of pain."

But then those authors go on to point out that pain also involves many regions of the brain:

> "The concept [of a pain center] is pure fiction unless virtually the whole brain is considered to be the 'pain center,' because the thalamus, the limbic system, the hypothalamus, the brain stem reticular formation, the parietal cortex, and the frontal cortex are all implicated in pain perception."

Perhaps we'll find more clues about how suffering works by studying a rare condition that results from injuring certain parts of the brain: the victims of *Pain Asymbolia* still recognize what the rest of us describe as pain—but do not find those feelings unpleasant, and may even laugh in response to them—which suggests that these patients have lost some resources that normally *cause* those cascades of torments.

In any case, to understand what suffering is, it won't suffice merely to learn where its machinery is; what we really need are better ideas about how those processes relate to our highest-level values, goals, and mental models of ourselves:

> Daniel Dennett 1978: "Real pain is bound up with the struggle to survive, with the real prospect of death, with the afflictions of our soft and fragile and warm flesh. . . . There can be no denying (though many have ignored it) that our concept of pain is inextricably bound up with (which may mean something less strong than essentially connected with) our ethical intuitions, our senses of suffering, obligation, and evil."

## Physical vs. Mental "Pain"

Are mental and physical pains the same? Suppose that you were to hear Charles say, *"I felt so anxious and upset that it felt as if something was tearing my gut."* You might conclude that Charles's feelings reminded him of times when he had a stomachache.

> Physiologist: It might even be true that your "stomach crawled"— if your mental condition caused your brain to send signals to your digestive tract.

Why do we so often speak as though "hurt feelings" resemble physical pains, although they have such different origins? Is there anything similar between the physical pain of a stomachache and the distress caused by disrespect from a friend? Yes, because, although these start with different kinds of events, being rejected by one's peers can eventually disrupt your brain in much the same way as can an abdominal pain.

> Student: As a child, I once hit my head on a chair, so I covered the injury with my hand. At first the pain was not intensive, but as soon as I noticed some blood on my hand, my suffering seemed to become much worse.

Presumably, the sight of blood does not affect the pain's intensity, but helps to engage higher-level activities. We undergo similar kinds of large-scale cascades in all sorts of situations like these:

> The grief of losing a long-term companion
> The helplessness of seeing others in pain
> The frustration of trying to stay awake
> The ache of humiliation or embarrassment
> The distraction that comes with excessive stress

## Feeling, Hurting, and Suffering

> "As he thought of it, a sharp pang of pain struck through him like a knife and made each delicate fiber of his nature quiver. His eyes deepened into amethyst, and across them came a mist of tears. He felt as if a hand of ice had been laid upon his heart."
> —*Oscar Wilde, in* The Picture of Dorian Gray

We have many words for types of Pain—like *stinging, throbbing, piercing, shooting, gnawing, burning, aching,* and so on. But words never capture quite enough of what any particular feeling *is,* so we have to resort to analogies that try to describe what each feeling is *like*—such as "a knife" or "a hand of ice"—or images of a suffering person's appearance. Dorian Gray felt no physical pain, but was horrified about growing old—hideous, wrinkled, and worst of all, of having his hair lose its beautiful gold.

*But what makes feelings so hard to describe?* Is this because feelings are so simple and basic that there's nothing more to be said about them? On the contrary, it seems to me that what we call "feelings" are what result from our attempts to describe our whole mental states—no matter that every such state is so complex that any brief description of it can capture just a few aspects of it. Consequently, the best we can do is to recognize some ways in which our present state is similar to or differs from some other states that we recollect. In other words, because our mental states are so complex, we can describe them only in terms of analogies.

Nevertheless, it can be easy to *recognize* (as opposed to *describe*) a particular feeling or mental condition because you may only need to detect a few of its characteristic features. This allows us to tell our friends enough about how we presently feel, because (assuming that both minds have somewhat similar structures) just a few clues may be enough for one person to recognize the other's condition. And in any case, most people know that this kind of communication or "empathy" is open to error as well as deception.

All this raises questions about what distinctions we are trying to make between what we call "Pain," "Hurting," and "Suffering." People sometimes use those terms as though they only distinguish intensities—but here I'll use "Pain" for sensations that arrive quickly after an injury, and use "Hurting" for what happens when this elevates the goal to get rid of the pain. Finally, I will use "Suffering" for the states that result when this escalates into a large-scale cascade that disrupts all one's usual Ways to Think.

> Philosopher: I agree that pain can lead to many kinds of changes in a person's mind, but that doesn't explain how suffering feels. Why can't all that machinery work without making people feel so bad?

It seems to me that when people talk about "feeling bad" they are referring to the disruption of their other goals, and to the various conditions that result from this. Pain would not serve the functions for which it evolved if it allowed us to keep pursuing our usual goals while our bodies were being destroyed. But if too much of the rest of the mind were suppressed, we might be unable to think of adequate ways to get rid of the pain—so we need to keep active some, at least, of our higher-level abilities. However, if we can still reflect on ourselves, then we are likely to get into the sorts of conditions called Remorse, Dismay, and Fear—all of which can be aspects of Suffering.

> Philosopher: Isn't there still something missing here? You have described a lot of processes that might be going on in our brains— but you have not said anything about why those conditions should give rise to any feelings at all. Why can't all that just happen without our having any sense of "experiencing" them?

Many philosophers have been puzzled by this mystery of why we have those "subjective experiences." I think that I have a good explanation for this, but it needs so many other ideas that we'll have to postpone it till Chapter 9.

## 3-4 Overriding Pain

> Sonja: "To love is to suffer. To avoid suffering one must not love.
> But then one suffers from not loving. Therefore, to love is to suf-
> fer; not to love is to suffer; to suffer is to suffer. To be happy is
> to love. To be happy, then, is to suffer, but suffering makes one
> unhappy. Therefore, to be happy one must love or love to suffer
> or suffer from too much happiness."
> —*Woody Allen, in* Love and Death

Some reactions to *Pain* are so brief that they're finished before one knows
that they are happening. If Joan happens to touch something hot, her arm
will quickly jerk her hand away before she's had time to think about it. But
Joan's reflexes cannot move her away from the pain in her knee because it
follows her everywhere she goes. By forcing one to focus on it, a persistent
pain can interfere with one's thinking of ways to get rid of it.

Of course, if Joan urgently wants to cross that room, she can probably
do it "in spite of the pain"—at the risk of further injury. Thus, professional
boxers and football players can train themselves to tolerate blows that are
likely to damage their bodies and brains. How do they manage to override
pain? We each know some methods for doing this and, depending on the
culture we're in, we regard some such techniques to be commendable but
others to be unacceptable.

> "About that time, G. Gordon Liddy began a new exercise in will
> power. He would burn his left arm with cigarettes, then matches
> and candles to train himself to overcome pain. . . . Years later,
> Liddy assured an acquaintance that he would never be forced to
> disclose anything he did not choose to reveal. He asked her to
> hold out a lit lighter. Liddy put his hand in the flame and held it
> there until the smell of burning flesh caused his friend to pull the
> flame away."
> —*Larry Taylor*[2]

If you keep your mind involved with other things, then a pain may seem
to feel less intense. We all have heard anecdotes in which a wounded sol-
dier continues to fight without being disrupted by pain—and only later
succumbs to shock, after the battle is lost or won. Thus, a powerful goal

to save yourself, or to save your friends, may be able to override everything else. On a smaller scale, with a milder pain, you may simply be too busy to notice it; then the pain may still "be there" but cannot get enough priority to disrupt your other activities.

Shakespeare reminds us (in *King Lear*) that misery loves company: no matter how awful one's lot may be, we still may draw comfort from knowing that the same could happen to someone else.

> When we our betters see bearing our woes,
> We scarcely think our miseries our foes.
> Who alone suffers suffers most i'th' mind,
> Leaving free things and happy shows behind;
> But then the mind much sufferance doth o'erskip
> When grief hath mates, and bearing fellowship.
> How light and portable my pain seems now,
> When that which makes me bend makes the King bow.

Another way to deal with pain is to apply a "counterirritant": when a certain part of your body aches, it sometimes helps to rub or pinch that spot—or to aggravate some different place. But why should a second disturbance offset the first, instead of simply making things worse? (See Melzack 1993.) A simple theory of this might be that when there are multiple sources of pain, it is hard for the rest of the brain to choose which of these sources to focus on. That could make it difficult for a single cascade to continue to grow.

Many other processes can alter how pain can affect our behavior:

> Aaron Sloman 1996: "Some mental states involve dispositions, which in particular contexts would be manifested in behavior, and if the relevant behavior does not occur then an explanation is needed (as with a person who is in pain not wincing or showing the pain or taking steps to reduce it). The explanation may be that he has recently joined some stoic-based religious cult, or that he wants to impress his girl friend, etc."

This applies to the treatment of pain-ridden people.

> Marian Osterweis 1987: "The degree of awareness of one's own pain may vary from a near denial of its presence to an almost total

preoccupation with it, and the reasons for attending to pain may vary. Pain itself may become the focus of the self and self-identity, or may, however uncomfortable, be viewed as tangential to person-hood. One of the most powerful influences on the way in which symptoms are perceived and the amount of attention paid to them is the meaning attributed to those symptoms."

Finally, in Chapter 9, we'll discuss the seeming paradox implied by the many common activities, such as in competitive sports, or in training for strength, in which one tries to do things beyond one's reach—because then, the greater the pain, then the higher the score.

## Prolonged and Chronic Suffering

When an injured joint becomes swollen and sore, and the slightest touch causes fiery pain, it's no accident that we call it "inflamed." As we noted in Section 3-1, this can be a benefit, by leading you to protect that site, thus helping that injury to heal. However, it is hard to defend the dreadful effects of those other, chronic pains that never end. Then we tend to ask questions like, *"What did I do to deserve this?"* Then if we can find some-thing that justifies punishment, it may bring us relief to be able to think, *"Now I can see why it serves me right!"*

Many victims discover no such escapes, and find that much has been lost from their lives; some even decide to end their lives. However, some others find ways to regard their sufferings as incentives or opportunities to show what they can accomplish, or even as unexpected gifts to help them to cleanse or renew their characters.

> F. M. Lewis 1982: "Becoming an invalid can be a blow to a person's self-esteem. However, for some patients, the sick role is seen as an elevation in status—deserving the nurturance and concern of others. The ability to assign meaning to an illness or to symptoms has been found to enhance some patients' sense of self-mastery over a problem or crisis."

Thus, some of those victims find ways to adapt to chronic, intractably painful conditions. They work out new ways to make themselves think and they rebuild their lives around those techniques. Here is how Oscar

Wilde describes how he dealt with the misery of his imprisonment in Reading Gaol:

> Wilde 1905: "Morality does not help me. I am one of those who are made for exceptions, not for laws. Religion does not help me. The faith that others give to what is unseen, I give to what one can touch, and look at. Reason does not help me. It tells me that the laws under which I am convicted, and the system under which I have suffered are wrong and unjust. But, somehow, I have got to make both of these things just and right to me. I have got to make everything that has happened to me good for me. The plank bed, the loathsome food, the hard ropes, the harsh orders, the dreadful dress that makes sorrow grotesque to look at, the silence, the solitude, the shame—each and all of these things I had to transform into a spiritual experience. There is not a single degradation of the body which I must not try and make into a spiritualizing of the soul."

Recent research on pain relief has developed new techniques, first for assessing degrees of pain and then for successfully treating it. We now have drugs that can sometimes suppress some of pain's most cruel effects—but many still never find relief, either by mental or medical means. It seems fair to complain that, in this realm, evolution has not done well for us— and this must frustrate theologians: *Why are people made to suffer so much? What functions could such suffering serve?*

Perhaps one answer to this is that the bad effects of chronic pain did not evolve from selection at all, but simply arose from a "programming bug." The cascades that we call "Suffering" must have evolved from earlier schemes that helped us to limit our injuries—by providing the goal of escaping from pain with an extremely high priority. The resulting disruption of other thoughts was only a small inconvenience before our ancestors evolved new, vaster intellects. In other words, our ancient reactions to chronic pains have not yet been adapted to be compatible with the reflective thoughts and farsighted plans that only later evolved in our brains. Evolution never had any sense of how a species might evolve next—so it did not anticipate how pain might disrupt our future high-level abilities. And thus, we came to evolve a design that protects our bodies but ruins our minds.

## Grief

> I cannot weep, for all my body's moisture
> Scarce serves to quench my furnace-burning heart;
> Nor can my tongue unload my heart's great burden,
> For self-same wind that I should speak withal
> Is kindling coals that fires all my breast,
> And burns me up with flames that tears would quench.
> To weep is to make less the depth of grief.
> Tears then for babes; blows and revenge for me!
> —*Shakespeare, in* Henry VI, Part 3

When you suffer the loss of a long-time friend, you feel that you've lost a part of yourself, because so many parts of your mind depend on that sharing of dreams and ideas—and now, alas, the signals that those brain parts transmit will never again receive replies. This is just like losing a hand or an eye—and that could be why it takes so much time to come to terms with being deprived of resources that you could rely on before that loss.

> *Gloucester:* Be patient, gentle Nell; forget this grief.
> *Duchess:* Ah, Gloucester, teach me to forget myself!
> —*Shakespeare, in* Henry VI, Part 2

Nell can't comply with Gloucester's advice because her links of affection are widely dispersed, rather than stored in some single place that she could select and then quickly erase. Besides, she may not *wish* to forget them all, as Aristotle suggests in *Rhetoric:*

> "Indeed, it is always the first sign of love, that besides enjoying someone's presence, we remember him when he is gone, and feel pain as well as pleasure, because he is there no longer. Similarly, there is an element of pleasure even in mourning and lamentation for the departed. There is grief, indeed, at his loss, but pleasure in remembering him and, as it were, seeing him before us in his deeds and in his life."

Here Shakespeare shows how we embrace our griefs and squeeze them till they take on pleasing shapes:

Grief fills the room up of my absent child,
Lies in his bed, walks up and down with me,
Puts on his pretty looks, repeats his words,
Remembers me of all his gracious parts,
Stuffs out his vacant garments with his form;
Then have I reason to be fond of grief.
—*Shakespeare, in* King John

## 3-5 **Mental Correctors, Suppressors, and Censors**

"Don't pay any attention to the critics. Don't even ignore them."
—*Sam Goldwyn*

Joan's sore knee has been getting worse. Now it hurts her all the time, even when it isn't touched. She thinks, "I shouldn't have tried to pick up that box. And I should have put ice on my knee at once."

It would be great never to make a mistake, or to get an idea that's not perfectly right—but we all make errors and oversights, not only in the physical realm but also in social and mental realms. However, although our decisions are frequently incorrect, it truly is remarkable how rarely these lead to catastrophes. Joan seldom sticks things in her eye. She scarcely ever walks into walls. She never tells strangers how ugly they are. How much of a person's competence is based on knowing which actions *not* to take?

We usually think of a person's abilities in positive terms, as in, *"An expert is someone who knows what to do."* But one could take the opposite view, that *"An expert is someone who rarely slips up—because of knowing what* not *to do."* However, this subject was rarely discussed in twentieth-century psychology—except, perhaps most notably, in Sigmund Freud's analyses.

Perhaps that neglect was inevitable, because, in the early 1900s, many psychologists became "behaviorists," who trained themselves to think only about the physical actions that people *do,* while ignoring questions about what people do *not* do. The result of this was to ignore what Chapter 6 will call *"negative expertise"*—which, I suspect, is a very large part of every person's precious collection of commonsense knowledge. In other words, much of what we come to know is based on learning from our mistakes.

To explain how our negative expertise works, I'll conjecture that our minds accumulate resources that we shall call *"Critics"—each of which*

*learns to recognize some particular kind of potential mistake.* I'll assume that
everyone possesses at least these three different types of Critics:

> A Corrector declares that you are doing something dangerous.
> *"You must stop right now, because you're moving your hand toward
> a flame."*

> A Suppressor interrupts before you begin the action you're plan-
> ning to take. *"Don't start to move your hand toward that flame, lest
> it get burned."*

> A Censor acts yet earlier, to prevent that idea from occurring to
> you—so you never even consider the option of moving your hand
> in that direction.

A Corrector's warning may come too late, because the action is already
going on; a Suppressor can stop it before it begins—but both can slow you
down by taking some time. In contrast, a Censor can actually speed you
up, by keeping you from considering the activities that it prohibits. This
could be one reason why experts are sometimes so quick; *they don't even
conceive of those wrong things to do.*

> Student: How could a Censor prevent you from thinking of
> something before you have started to think about it? Isn't that
> some kind of paradox?

> Programmer: No problem. Design each Censor to be a machine
> that is equipped with enough memory that it can remember the
> way you were thinking several steps before you made a certain
> particular kind of mistake. Then later, when that Censor recog-
> nizes a similar state, it steers you to think in some different way so
> that you then won't repeat that mistake.

Of course, excessive cautiousness could have bad effects. If your Critics
tried to prevent you from making every conceivable type of mistake, you
might become so conservative that you would never try to do anything
new. You might never be able to cross a street, because you could always
conceive of some way you could meet with some accident. On the other

side, it would be dangerous to not have enough Critics, because then you would make too many mistakes. So here we'll briefly talk about what might happen when we switch between these two extremes.

## What Happens When Too Many Critics Get Switched?

> I have of late—but wherefore I know not—lost all my mirth, forgone all custom of exercises; and indeed it goes so heavily with my disposition, that this goodly frame, the earth, seems to me a sterile promontory; this most excellent canopy, the air, look you, this brave o'erhanging firmament, this majestical roof fretted with golden fire, why, it appeareth nothing to me but a foul and pestilent congregation of vapors.
>
> —*Shakespeare, in* Hamlet

In later chapters we'll argue that much of our human resourcefulness comes from our ability to switch among different Ways to Think. However, this could also be the source of many of the conditions we call our tempers, moods, and dispositions—as well as our many and varied mental disorders. For example, if certain Critics were to stay active all the time, then one would appear to be obsessed with certain aspects of the world or oneself—or else one might constantly seem to be compelled to repeat certain kinds of activities. Another example of poor Critic control would be when one repeatedly turns too many Critics on, and later switches too many off. Here is what appears to be a firsthand description of such a condition:

> Kay Redfield Jamison 1994: "The clinical reality of manic-depressive illness is far more lethal and infinitely more complex than the current psychiatric nomenclature, bipolar disorder, would suggest. Cycles of fluctuating moods and energy levels serve as a background to constantly changing thoughts, behaviors, and feelings. The illness encompasses the extremes of human experience. Thinking can range from florid psychosis, or "madness," to patterns of unusually clear, fast and creative associations, to retardation so profound that no meaningful mental activity can occur. Behavior can be frenzied, expansive, bizarre, and seductive, or it can be seclusive, sluggish, and dangerously suicidal. Moods

may swing erratically between euphoria and despair or irritability and desperation. . . . [But] the highs associated with mania are generally only pleasant and productive during the earlier, milder stages."

A later paper by Jamison goes on to suggest that some value can come from those massive cascades:

Kay Redfield Jamison 1995: "It seems, then, that both the quantity and quality of thoughts build during hypomania. This speed increase may range from a very mild quickening to complete psychotic incoherence. It is not yet clear what causes this qualitative change in mental processing. Nevertheless, this altered cognitive state may well facilitate the formation of unique ideas and associations. . . . Where depression questions, ruminates and hesitates, mania answers with vigor and certainty. The constant transitions in and out of constricted and then expansive thoughts, subdued and then violent responses, grim and then ebullient moods, withdrawn and then outgoing stances, cold and then fiery states—and the rapidity and fluidity of moves through such contrasting experiences—can be painful and confusing.

It is easy to recognize such extremes in the mental illnesses called "bipolar" disorders, but I suspect that everyone constantly uses such processes in the course of their everyday commonsense thinking! Thus, Chapter 7 will suggest that, whenever you face a new type of problem, you might find solutions by using procedures like this:

First, briefly shut most of your Critics off. This helps you to think of some things you could do—with little concern about whether they'll work—as though you were in a brief "manic" state.

Next, turn many Critics on, to examine these options more skeptically—as though you were having a mild depression.

Finally, choose an option that seems promising, and then proceed to pursue it, until one of your Critics starts to complain that you have stopped making progress.

Sometimes you may go though such phases deliberately, perhaps spending several minutes on each. However, my conjecture is that we often do this on timescales of one or two seconds, or less, in the course of our everyday commonsense thinking. But then, all these events may be so brief that we have almost no sense that they're happening.

## The "Critic-Selector Model of Mind"

Chapter 1 described an animal as little more than a system based on a catalog of *If→Do* rules, where each *If* describes a type of physical situation, and its *Do* describes a useful way to react to it.

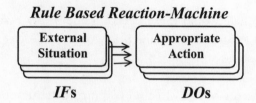

Chapter 7 will extend this to what I will call the *"Critic-Selector Model of Mind,"* which portrays our thinking as based on *mental reactions* to *mental situations.* In this model, our Critics play a central role in making large-scale changes in how we think, by selecting resources we'll use for thinking about different kinds of situations. Here is a simplified version of this:

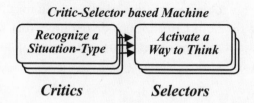

Each of these Critics learns to recognize some particular kind of mental condition so that whenever that condition occurs, this Critic will try to activate one or more *sets of resources* that have been useful, in the past, for dealing with that type of mental situation.

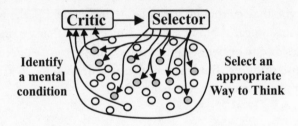

**A CRITIC SELECTING A SET OF RESOURCES**

Chapter 7-3 will suggest more ideas about how these resources are formed and organized.

> Student: Where would those Critics reside in my brain? Would they all be located in the same place, or would each part of the brain have some of its own?

Our Critic-Selector Model of Mind will include structures like these at every level, so that each person's mind will include reactive, deliberative, and reflective Critics. At the lowest levels, those Critics and Selectors are almost the same as the *Ifs* and *Then*s of simple reactions. But at our higher reflective levels, these Critics and Selectors can cause so many changes that, in effect, they switch us to different Ways to Think. (See Singh 2003b.)

I should note that the word *Critic* is often restricted to mean a person who only detects deficiencies. However, it also is useful to recognize when a strategy works better than we expected—and then to bestow more priority, time, or energy to the process that deserves credit for this. So, Chapter 7-2 will extend the term *Critic* to include resources that not only detect mistakes but also recognize successes and promising opportunities; we'll call those "positive" critics "*Encouragers.*"

## 3-6 The Freudian Sandwich

> Luck's a chance, but trouble's sure,
> I'd face it as a wise man would,
> and train for ill and not for good.
> —*A. E. Housman*

Few textbooks of psychology discuss how we decide what *not* to think about. However, this was a major concern to Sigmund Freud, who envisioned the mind as a system in which ideas need to overcome barriers.

> Sigmund Freud 1920: "[The mind includes] a large anteroom in which the various mental excitations are crowding upon one another, like individual beings. Adjoining this is a second, smaller apartment, a sort of reception-room, in which consciousness resides. But on the threshold between the two there stands a personage with the office of doorkeeper, who examines the various mental excitations, censors them, and denies them admittance to the reception-room when he disapproves of them. You will see at once that it does not make much difference whether the doorkeeper turns any one impulse back at the threshold, or drives it out again once it has entered the reception-room. That is merely a matter of the degree of his vigilance and promptness in recognition."

However, getting past this first barrier is not quite enough to make us reflect on a possible thought—or what Freud calls a mental excitation—because, as he goes on to say, this leads only to the reception room:

> "The excitations in the unconscious, in the antechamber, are not visible to consciousness (which is in the other room), so, to begin with, they remain unconscious. When they have pressed forward to the threshold and been turned back by the doorkeeper, they are 'incapable of becoming conscious'; we call them then repressed. But even those excitations which are allowed over the threshold do not necessarily become conscious; they can only become so if they succeed in attracting the eye of consciousness."

Thus, Freud imagined the mind as an obstacle course in which only ideas that get far enough are awarded the status of consciousness. In one kind of block (which Freud calls "repression"), an impulse is blocked at an early stage—without the thinker becoming aware of this. However, repressed ideas can still persist—and may be expressed in elusive disguises—by changing the manner in which they're described (so that the Censors no longer can recognize them). Freud used the term *sublimation* for this, but we sometimes call this "rationalizing." Finally, an idea can reach the highest level and still be rendered powerless, although one can remember rejecting it (Freud names this process "repudiation.")

More generally, Freud suggests that the human mind is like a battleground in which many resources are working at once—but don't always share the same purposes. Instead, there often are serious conflicts between our animal instincts and our acquired ideals. Then the rest of the mind must either find ways to compromise or else to suppress some of those competitors.

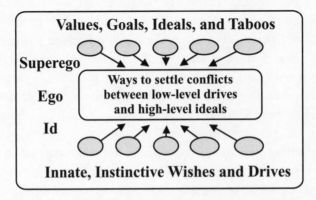

THE FREUDIAN SANDWICH

It is more than a century since Sigmund Freud recognized that human thinking does not proceed in any single, uniform way. Instead, he saw each mind as a host of diverse activities that often lead to conflicts and inconsistencies—and he saw that our various ways to deal with these involve many different processes, which in everyday life we try to describe with vague suitcase-like names such as *Conscience, Emotion,* and *Consciousness.*

## 3-7 Controlling Our Moods and Dispositions

"Love, he believed, made a fool of a man, and his present emotion was not folly but wisdom; wisdom sound, serene, well-directed. . . . She seemed to him so felicitous a product of nature and circumstance that his invention, musing on future combinations,

was constantly catching its breath with the fear of stumbling into some brutal compression or mutilation of her beautiful personal harmony. . . . "
—*Henry James, in* The American

Chapter 1-2 described how our feelings and attitudes frequently swing between extremes:

"Sometimes a person gets into a state where everything seems to be cheerful and bright—although nothing outside has actually changed. Other times everything pleases you less: the entire world seems dreary and dark, and your friends complain that you seem depressed."

We use terms like *dispositions* and *moods* to refer to these kinds of conditions, in which we change the subjects we think about, and the ways in which we think about them. At first one may think about physical things, then about some social matters, and then one may start to reflect on one's longer-term goals and plans. But what determines the length of time that a person will stay in each such frame of mind, before switching to some other concern?

A flash of anger, fear, or a sexual image may last for only an instant, while other moods can last for minutes or hours—and some may continue for weeks or years. "John is angry" means that he's angry now—but "an angry kind of person" may describe a lifelong trait. On what do these durations depend? Perhaps this partly depends on how our mental Critics are managed.

Clearly, some of our Critics are always on the job, like voyeurs that constantly monitor us, waiting for moments to set off alarms—whereas other Critics are only active on special occasions, or in particular states of mind. Let's look again at two extremes:

If you could switch all your Critics off, then nothing would seem to have any faults, and the whole world might suddenly seem to change so that everything now seems glorious. You'd be left with few worries, concerns, or goals—and others might describe you as elated, euphoric, demented, or manic.

However, if you turned too many Critics on, you'd see imperfections everywhere. Your entire world would seem filled with flaws, engulfed in a flood of ugliness. If you also found fault with your goals themselves, you'd feel no urge to straighten things out, or to respond to any encouragement.

This means that those Critics must be controlled: if you were to turn too many on, then you would never get anything done. But if you turned all your Critics off, it might seem as though all your goals were achieved—and again you wouldn't accomplish much.

So let's look more closely at what could happen if something turned most of your Critics off. If you'd like to experience this yourself, there are some well-known steps you can take.[3] It would help to be suffering pain and stress, so starvation and cold may also assist—so will psychoactive drugs. Meditation can have such effects, and it helps to move into some strange, quiet place. Next, you could set up a rhythmical drone that repeats some monotonous phrase or tone, and soon it will lose all meaning and sense—and so will virtually everything else! Then, if you can get yourself into such a condition, you'll have a chance to undergo this singular kind of experience:

> Meditator: It suddenly seemed as if I was surrounded by an immensely powerful Presence. I felt that a Truth had been "revealed" to me that was far more important than anything else, and for which I needed no further evidence. But when later I tried to describe this to my friends, I found that I had nothing to say except how wonderful that experience was.

This peculiar type of mental state is sometimes called a "Mystical Experience" or "Rapture," "Ecstasy," or "Bliss." Some who undergo it call it "wonderful," but a better word might be "wonderless," because I suspect that such a state of mind may result from turning so many Critics off that one cannot find any flaws in it.

What might that "powerful Presence" represent? It is sometimes seen as a deity, but I suspect that it is likely to be a version of some early Imprimer that for years has been hiding inside your mind. In any case, such experiences can be dangerous—for some victims find them so compelling that they devote the rest of their lives to trying to get themselves back to that state again.

Nevertheless, in everyday life there remains a wide range in which it's both useful and safe to regulate your collection of Critics. Sometimes you feel adventurous, inclined to try new experiments. Other times you feel conservative, and try to avoid uncertainty. And when you're in an emergency and don't have time to reason things out, you may need to set aside your long-range plans and expose yourself to pain and stress. To do this, you'll have to suppress at least some of your Correctors and Censors.

All this raises many questions about how we develop our mental Critics. How do we make them and how do we change them? Do some of them scold other Critics when they produce poor performances? Are certain minds more productive because their Critics are better organized? We'll come back to such questions in Chapter 7-6.

## 3-8 Emotional Exploitation

Whatever you may be trying to do, your brain may have other plans for you.

> I was trying to work on a difficult problem, but was beginning to fall asleep. Then I found myself imagining that my friend Professor Challenger was about to develop the same technique. This caused a flicker of angry frustration, which blocked for the moment my urge to sleep—and this allowed me to finish my job.[4]

In fact, Challenger was not doing any such thing; he works in a totally different field—but we had recently been in an argument, so he could serve as a person to be angry at. Let's make up a theory of how this worked.

> A resource called Work was attending to one of my goals, but the process called Sleep tried to take over control. Then, somehow, I constructed that fantasy—and the resulting annoyance and jealousy counteracted that urge to sleep.

All of us use this kind of trick to combat frustration, boredom, hunger, or sleep. By self-inducing anger or shame, you can sometimes counteract fatigue or pain—as when one is falling behind in a race, or trying to lift too heavy a weight. With such emotional "double negatives," you can use one system to switch off another. However, such "self-control" tactics must be used cautiously. If you don't make yourself angry enough, you might relapse into lassitude—whereas if you make yourself become too irate, you might completely forget what you had wanted to do. Sometimes, just a tweak of anger might ward off sleep, in a moment so brief that you don't notice it.

Here's another example where part of a mind "exploits" one emotion for the purpose of turning off another—thus helping you to attain some goal that you cannot achieve more directly.

> Joan is trying to follow a diet. When she sees that chocolate cake, she is filled with a strong temptation to eat. But when she imagines a certain friend looking gorgeous in her bathing suit, then Joan's craving to have a similar shape keeps her from actually eating the cake.

How might such a fantasy work to produce that kind of effect? Joan has no straightforward way to suppress her reckless appetite—but she knows that the sight of her rival makes her more concerned about her body's shape. Therefore, arousing that image is likely to diminish her urge to eat. (Of course, that strategy carries some risk: if her jealousy makes Joan feel depressed, she might engorge the entire cake.)

> Citizen: Why should we need to use fantasies to induce ourselves to do such things when we know that those images aren't real? Why can't we use more rational ways to figure out what we should do?

One answer is that the concept of "rational" itself is a kind of fantasy—because our thinking is never entirely based on purely logical reasoning. To us, it might seem "irrational" to exploit an emotion to solve a problem. However, when Joan's *Losing Weight* encounters an obstacle, it makes just as much sense for that goal to exploit emotions like *Jealousy* or *Disgust* as it would for Joan herself to use a stick to extend her reach—no matter that even Joan herself may see such behaviors as "emotional."

Besides, we're always exploiting fantasies in the course of our everyday commonsense thinking. When you sit at a table across from friends, you cannot see their backs or legs, but this is of no concern to you because most of what you think you see comes from your internal models and memories. For while some parts of your brain get information from the outer world, most of them are reacting to information they get from other processes inside your brain. Indeed, a major part of our daily lives consists of imagining things we don't have but might need—such as a forthcoming vacation. *More generally, to think about changing the way things are, we have to imagine how they might be.*

> Citizen: I agree that we frequently do such things—but why should we need to tell lies to ourselves? Why can't we directly just turn off Sleep, instead of resorting to fantasies? Why can't we simply command our minds to do whatever we want them to do?

One answer seems clear: *directness would be too dangerous.* If some other goal could simply turn Hunger off, we'd all be in peril of starving to death. If it could directly switch Anger on, we might find ourselves fighting most of the time. If it could simply extinguish Sleep, we'd be likely to wear our bodies out. So this shaped the way our brains evolved the instinctive reactions that keep us alive—by making it hard to hold one's breath, or to keep from falling asleep, or to control how much you eat; those who were able to do such dangerous things left fewer descendants than did the rest.

# 4
# CONSCIOUSNESS

## 4-1  What in the World Is Consciousness?

"No philosopher and hardly any novelist has ever managed to explain what that weird stuff, human consciousness, is really made of. Body, external objects, darty memories, warm fantasies, other minds, guilt, fear, hesitation, lies, glees, doles, breath-taking pains, a thousand things which words can only fumble at, co-exist, many fused together in a single unit of consciousness."
  —*Iris Murdoch, in* The Black Prince

What kinds of creatures have consciousness? Does it exist in chimpanzees—or in gorillas, baboons, or orangutans? What about dolphins or elephants? Are crocodiles, frogs, or fish aware of themselves to any extent—or is consciousness a singular trait that distinguishes us from the rest of the beasts?

Of course, those animals won't answer questions like, "*What is your view of the nature of mind.*" But when we interview mystical thinkers who claim to know what consciousness is, their replies are seldom more enlightening.

Sri Chinmoy 2003: "Consciousness is the inner spark or inner link in us, the golden link within us that connects our highest and most illumined part with our lowest and most unillumined part."

Some philosophers even insist that no one has better ideas about this.

> Jerry Fodor 1992: "Nobody has the slightest idea how anything
> material could be conscious. Nobody even knows what it would be
> like to have the slightest idea about how anything material could be
> conscious. So much for the philosophy of consciousness."

Is consciousness an all-or-none trait that has a clear and definite boundary?

> Absolutist: We don't know where consciousness starts and stops,
> but every object must be conscious or not—and, clearly, people
> are conscious, while rocks are not.

Or does consciousness come in different degrees?

> Relativist: Everything has some consciousness. An atom has only
> a little of it, while brains can have it to greater degrees—and per-
> haps there are no limits to it.

Or is that question still too vague to justify trying to answer it?

> Logicist: Before you go on about consciousness, you really ought
> to define it. Good arguments should start right out by stating pre-
> cisely what they are about. Otherwise, you'll begin with a shaky
> foundation.

The Logicist's policy might seem "logical"—but, although we don't like to
be imprecise, *a clear definition can make things worse,* until we're sure that
our ideas are right. For, *consciousness* is one of those suitcase-like words that
we use for many types of processes, and for different kinds of purposes.
It's the same for most of our other words about minds, such as *awareness,
sentience,* or *intelligence.*[1]

So instead of asking what consciousness *is,* we'll try to examine when,
how, and why people use those mysterious words. But why do such questions even arise? What, for that matter, are *mysteries?*

> Daniel Dennett 1991: "A mystery is a phenomenon that people
> don't know how to think about—yet. Human consciousness is just

about the last surviving mystery. There have been other great mysteries [like those] of the origin of the universe and of time, space, and gravity. . . . However, consciousness stands alone today as a topic that often leaves even the most sophisticated thinkers tongue-tied and confused. And, as with all of the earlier mysteries, there are many who insist—and hope—that there will never be a demystification of consciousness."

Indeed, many of those who "insist—and hope" that consciousness cannot be explained still maintain that it alone is the source of most of the virtues of human minds.

Thinker 1: Consciousness is what binds all our mental events together, and thus unifies our present, past, and future into our continuous sense of experience.

Thinker 2: Consciousness makes us "aware" of ourselves, and gives us our sense of identity; it is what animates our minds and gives us our sense of being alive.

Thinker 3: Consciousness is what gives things meaning to us; without it, we would not even know we had feelings.

Wow! Wouldn't it be astonishing if any one principle, power, or force could endow us with all those abilities?

However, I'll argue that it would be a mistake to believe in any such entity—because we ought to be asking this question, instead: *"Isn't it remarkable that any single word or phrase could have come to mean so many different things?"*

William Calvin and George Ojeman 1994: "Modern discussions of consciousness . . . usually include such aspects of mental life as focusing your attention, things that you didn't know you knew, mental rehearsal, imagery, thinking, decision making, awareness, altered states of consciousness, voluntary actions, subliminal priming, the development of the concept of self in children, and the narratives we tell ourselves when awake or dreaming."

All this should lead us to conclude that *consciousness* is a suitcase-like word that we use to refer to many different mental activities, which don't have a single cause or origin—and, surely, this is why people have found it so hard to "understand what consciousness is." The trouble was that they tried to pack into a single box all the products of many processes that go on in different parts of our brains—and this produced a problem that will remain unsolvable until we find ways to chop it up. However, once we imagine a mind as made of smaller parts, we can replace that single, big problem by many smaller, more solvable ones—which is just what this chapter will try to do.

## 4-2 Unpacking the Suitcase of Consciousness

> Aaron Sloman 1994: "It is *not* worth asking how to define con-sciousness, how to explain it, how it evolved, what its function is, etc., because there's no one thing for which all the answers would be the same. Instead, we have many sub-capabilities, for which the answers are different: e.g., different kinds of perception, learning, knowledge, attention control, self-monitoring, self-control, etc."

To see the variety of what human minds do, consider this fragment of everyday thinking.

> *Joan is starting to cross the street on the way to deliver her finished report. While thinking about what to say at the meeting, she hears a sound and turns her head—and sees a quickly oncoming car. Uncertain as to whether to cross or retreat, but uneasy about arriving late, Joan decides to sprint across the road. She later remembers her injured knee and reflects upon her impulsive decision. "If my knee had failed, I could have been killed. Then what would my friends have thought of me?"*

It might seem natural to ask, *"How conscious was Joan of what she did?"* But rather than dwell on that *consciousness* word, let's look at some things that Joan actually *did*.

> Reaction: Joan reacted quickly to that sound.
> Identification: She recognized it as being a sound.

Specification: She classified it as the sound of a car.
Attention: She noticed certain things rather than others.
Indecision: She wondered whether to cross or retreat.
Imagining: She envisioned two possible future conditions.
Selection: She selected a way to choose among options.
Decision: She chose one of several alternative actions.
Planning: She constructed a multistep action plan.
Reconsideration: Later she reconsidered this choice.

She also did other things like these.

Learning: She created descriptions and stored them away.
Recollecting: She retrieved descriptions of prior events.
Embodiment: She tried to describe her body's condition.
Expression: She constructed some verbal representations.
Narration: She arranged these into storylike structures.
Intention: She changed some goals and priorities.
Apprehension: She was uneasy about arriving late.
Reasoning: She made various kinds of inferences.

She also used many processes that involved reflecting on what some of those other processes did.

Reflection: She thought about what she had recently done.
Self-Reflection: She reflected on what she had thought about.
Empathy: She imagined some other persons' thoughts.
Reformulation: She revised some of her representations.
Moral Reflection: She evaluated what she has done.
Self-Awareness: She characterized her mental condition.
Self-Imaging: She made and used models of herself.
Sense of Identity: She regarded herself as an entity.

This is only the start of a catalog of some of Joan's mental activities—and if we want to understand how her thinking works, we'll need to have much better ideas about how each of those activities work and how they all are organized. At various points in the rest of this book, we'll examine each item on that list and try to break it into parts—to see what processes it might involve. However, to accomplish this, we'll need to begin with some way or ways to divide an entire mind into parts—and our everyday folk-psychology abounds with ideas about dividing the functions of minds into pairs like these:

Conscious vs. Unconscious
Premeditated vs. Impulsive
Deliberate vs. Spontaneous
Intentional vs. Involuntary
Cognitive vs. Subcognitive[2]

We'll discuss such "dumbbell" distinctions in Chapter 9-2, and will conclude that each such division is simply too crude. For example, the division between *conscious* and *unconscious* does not distinguish between information that is inaccessible because one has no way to access it, or because it is actively censored or "repressed," or because (as Freud suggested) it has been "sublimated" into some form that one cannot recognize—or because one has simply failed to retrieve it (that is, to bring it into one's active working memory). In any case, this book will argue that little good will come from attempts to divide our minds into only two parts.

We have already seen some useful ways to split a mind into large numbers of different parts—for example, as sets of resources or as collections of rules. However, for making better generalizations, we'll need a design that has fewer components. Accordingly, every chapter of this book will exploit the idea that a mind is composed of processes that operate on just a few "levels." Beginning with three such levels will help us to avoid "dumbbell" distinctions, and the following chapter will argue that we'll need at least three more higher levels of mind. Nevertheless, the rest of this chapter will mainly focus on the question of why people are so prone to pack so many different concepts into that single "suitcase of consciousness."

## 4-3 *A*-Brains and *B*-Brains

*Socrates:* Imagine men living in an underground den, which has an opening towards the light—but the men have been chained from their childhood so that they never can turn their heads around and can only look toward the back of the cave. Far behind them, outside the cave, a fire is blazing, and between the fire and the

prisoners there is a low wall built along the way, like the screen
that puppeteers have in front of them, over which they show the
puppets.

*Glaucon:* I see.

*Socrates:* And do you see men passing along the wall carrying all sorts of vessels, and statues and figures of animals made of
wood, stone, and various materials, which appear over the wall?
Some of them are talking, others silent.

*Glaucon:* You have shown me a strange image. . . .

*Socrates:* Like us, they see nothing but only the shadows of
themselves and of those other objects, which the fire throws on
the opposite wall of the cave. . . . Then in every way such prisoners
would deem reality to be nothing else than those shadows. . . .

—*Plato, in* The Republic

Can you think about what you are *thinking right now*? In a literal sense,
that's impossible—because each new such thought would alter the thoughts
that you were just thinking before. However, you can settle for something
slightly less, by imagining that your brain (or mind) is composed of two
principal parts: Let's call these your *"A-Brain"* and *"B-Brain."*

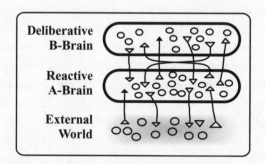

Now suppose that your *A*-Brain gets some signals from the external
world (via such organs as eyes, ears, nose, and skin)—and that it also can
react to these by sending signals that make your muscles move. By itself,
the *A*-Brain is a separate animal that only reacts to external events but has
no sense of what they might mean. For example, when the fingertips of
two lovers come into intimate physical contact, the resulting sensations,
by themselves, have no particular implications. For there is no significance

in those signals themselves: their meanings to those lovers lie in how they represent and process them in the higher levels of their minds. (See Pohl 1970.)

Similarly, your *B*-Brain is connected so that it can react to signals that it receives from *A,* and then can react by sending signals to *A.* However, *B* has no direct connection to the outer world, so, like the prisoners in Plato's cave, who see only shadows on a wall, the *B*-Brain mistakes *A*'s descriptions for real things. The *B*-Brain does not realize that what it perceives are not objects in the external world but are merely events in the *A*-Brain itself.

> Neurologist: That also applies to you and me. For whatever you think you touch or see, the higher levels of your brain never can actually contact these—but can only interpret the representations of them that your mental resources construct for you.

Nevertheless, although the *B*-Brain cannot directly perform any physical actions, it still can affect the external world, by controlling the ways in which *A* might react. For example, if *B* sees that *A* has got stuck at repeating itself, it might suffice for *B* to instruct *A* to change its strategy.

> Student: Sometimes, when I've misplaced my eyeglasses, I keep looking for them in the very same place. Then a silent voice reproaches me, suggesting that I stop repeating myself. But what if I were crossing a street when suddenly my *B*-Brain said "Sir, you've repeated the same actions with your leg for more than a dozen consecutive times. You should stop right now and do something else." That could cause me a serious accident.

To prevent such mistakes, a *B*-Brain would need appropriate ways to represent things. In this case, you would be better off if your *B*-Brain represented *"walking to a certain place"* as a single extended act, like, *"Keep moving your legs till you get to the other side of the street."*

However, this raises the question of how that *B*-Brain could acquire such skills.[3] Some could be built into it from the start, but, for the *B*-Brain to learn new techniques, it might itself need similar help, which could come from a level above it. Then while the *B*-Brain deals with its *A*-Brain world, that "*C*-Brain" in turn will supervise *B*.

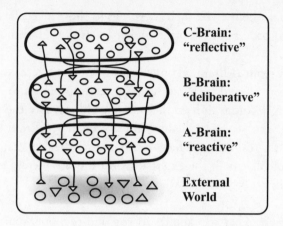

Student: Would not this raise increasingly difficult questions, because each higher level would need to be smarter and wiser?

Not necessarily, because that *C*-Brain could act like a "manager" who has no special expertise about how to do any particular job—but still could give "general" guidance like this:

If *B*'s descriptions seem too vague, *C* tells it to use more specific details.
If *B*'s are buried in too much detail, *C* suggests more abstract descriptions.
If what *B* is doing is taking too long, *C* tells it to try some other technique.

Furthermore, if both *B*-Brain and *C*-Brain get stuck, we could add yet more levels to our multilayer mind-machine.

Student: How many such levels does a person need? Do we have dozens or hundreds of them?

## Levels, Layers, and Organisms

This book suggests many reasons to think that our human mental resources are organized into at least these six levels of processes, as illustrated in the next figure:

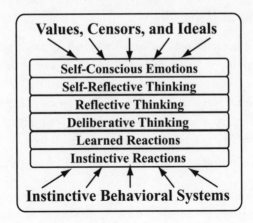

We can see each of these as aspects of Joan's decision to hurry across that street:

> What caused Joan to turn toward that sound? [Instinctive]
> How did she know that it was the sound of a car? [Learned]
> What resources were used to make her decision? [Deliberative]
> How did she choose which resources to use? [Reflective]
> Did she feel that she made a good decision? [Self-Reflective]
> Did her actions live up to her principles? [Self-Conscious]

We know that by the time of birth, every infant is already equipped with a variety of *instinctive* reactions, and has started to add *learned* reactions to these. Then, over time, we progressively add more *deliberative* ways to reason, imagine, and plan for the future. Later, we build a new layer in which we start to do *reflective thinking* about our own thoughts—and two-year-old children already are making additional ways to *self-reflect* about why and how they thought those things. And, eventually, we begin to think more *self-consciously* about which things to regard as *right* or *wrong* to do. Chapter 5 will add more details about how such systems might be organized.

> Student: Does your theory really need so many different levels? Are you sure that you can't make do with fewer of them? Indeed, why should we need any "levels" at all—instead of a single big, cross-connected network of resources?

## The Evolution of Psychology

There is an evolutionary reason for why we should not expect the brain to be a single, highly interconnected network: it would be almost impossible for such a system to evolve—because it would have so many flaws or "bugs" that it could not survive for long. And, of course, no system could do very much if its parts were not interconnected enough. This means that *whenever we increase a system's size, its performance is likely to decline—unless we also improve its design.* Let's give this argument a name:

> *The Organism Principle:* When a system evolves to become more complex, this always involves a compromise: if its parts become too separate, then the system's abilities will be limited—but if there are too many interconnections, then each change in one part will disrupt many others.

This surely is the reason why the bodies of all living things are composed of the distinctively separate parts we call "organs." In fact, that's why we call them "organisms":

> Organism: A body made up of organs, organelles, or other parts that work together to carry on the various processes of life.

This also applies to the organs called brains:

> Embryologist: In its early development, a typical structure in the brain starts out with more or less definite layers or levels like those in your A, B, C diagrams. But later those layers become less well defined because various groups of cells grow connections to other, more distant locations.

During the eons through which our brains evolved, our ancestors had to adapt to thousands of different environments—and, during each such episode, some structures that worked well in earlier times now behaved in some dangerous ways, so we had to evolve corrections for them. However, the evolution of a species is also constrained by the fact that it is extremely dangerous to make any change in the earlier stages of an animal's development—*because most of the structures that later evolved depend very much on how those earlier structures work.*

Consequently, evolution often works by adding *new fixes and patches that modify structures that have been already established.* For example, after certain major stages of growth in the brain, many new cells are later destroyed by "post-editing" processes that evolved to delete some types of connections.

The same sort of constraint also seems to apply whenever we try to improve the performance of any large system. For example, after every change we make in an existing computer program, we usually find that this has created additional bugs—and then we need to make yet more corrections. In fact, many computer systems eventually become so ponderous that their further development stops, because their programmers can no longer keep track of what all the previous programmers did.

Similarly, it appears that our brains result from processes in which each new part in based on some older designs, but also includes exceptions to it. Indeed, I suspect that large parts of our brains work mainly to correct mistakes that other parts make—and this is surely one reason why the subject of human psychology has become so hard. We can expect to discover neat rules and laws that partly explain many aspects of how we think. However, every such "law of thought" will also need a sizable list of exceptions to it. So psychology will never be much like physics, in which we frequently find "unified theories" that work flawlessly.

## Why Can't We See How Our Own Minds Work?

Why cannot we simply look into our minds to see precisely how they work? Why can't minds completely inspect themselves? Whatever those limitations may be, the philosopher Hume concluded that we could never surmount all of them:

> David Hume 1748: "The motion of our body follows upon the command of our will. Of this we are every moment conscious. But the means, by which this is effected; the energy, by which the will performs so extraordinary an operation; of this we are so far from being immediately conscious, that it must for ever escape our most diligent enquiry."

I suspect that Hume was right to think that no mind could wholly understand itself by trying to look inside itself. One problem is that each part of the brain does much of its work in ways that other parts cannot observe.

Another obstacle is that when any part tries to examine another, that probing may alter the state of that other part, thus corrupting the very evidence that the first part was trying to get.

However, way back in 1748, not even David Hume could predict that we would develop instruments that could look inside a living brain without destroying any evidence. Yet today, every year brings new scanning machines that reveal even more details of the processes that we call "mental events." Nevertheless, some thinkers still claim that this will never tell us enough:

> Dualist philosopher: All such methods are doomed to fail because, although you can measure or weigh the parts of a brain, no physical instrument can ever detect subjective experiences like thoughts or ideas, which exist in a separate mental world.

Such thinkers believe that our feelings are caused by nonphysical processes that will forever remain beyond the realm of scientific explanations. However, I'll argue that this opinion results from squeezing too many different questions into a single word like "subjective." That gives us the illusion we're facing one single, unsolvable mystery—but Chapter 9 will try to show that, although some of those questions are difficult, we can make progress on all of them by dealing with each of them separately.

> Holist: I don't believe that approach will work because consciousness is just one of those "wholes" that emerges inexplicably whenever a system gets complex enough. And that is just what we should expect from the network of billions of cells in a brain.

If mere complexity were enough, then almost everything would have consciousness! For example, the manner in which a wave breaks on a beach is more complex in most respects than the processes that go on in a brain— but this should not lead us to conclude that waves think. For as our Organism Principle says, if a system's parts have too many connections, nothing but traffic jams will "emerge"—while if its interconnections are too sparse, then the system will do almost nothing at all.

All these arguments suggest that there is little to gain from wondering what consciousness "is"—because that word includes too much for us to deal with all at once. Let's listen to Aaron Sloman again:

Aaron Sloman 1992: "I, for one, do not think defining consciousness is important at all, and I believe that it diverts attention from important and difficult problems. The whole idea is based on a fundamental misconception that just because there is a noun 'consciousness' there is some 'thing' like magnetism or electricity or pressure or temperature, and that it's worth looking for correlates of that thing. Or on the misconception that it is worth trying to prove that certain mechanisms can or cannot produce 'it,' or trying to find out how 'it' evolved, or trying to find out which animals have 'it,' or trying to decide at which moment 'it' starts when a fetus develops, or at which moment 'it' stops when brain death occurs, etc. There will not be one thing to be correlated but a very large collection of very different things."

I completely agree with Sloman's view. To understand how our thinking works, we must study each of those "very different things" and then ask what kinds of machinery could accomplish some or all of them. *In other words, we must try to* design—*as opposed to* define—*machines that can do what human minds do.*

## 4-4 Overrating Consciousness

Wilhelm Wundt 1897: "Our mind is so fortunately equipped that it brings us the most important bases for our thoughts without our having the least knowledge of this work of elaboration. Only the results of it become conscious. This unconscious mind is for us like an unknown being who creates and produces for us, and finally throws the ripe fruits in our lap."

One reason why consciousness seems so mysterious is that we exaggerate our perceptiveness. For example, as soon as you enter a room, you have the sense that you instantly see everything that is in your view. However, this is far from true: it is an illusion that comes because your eyes so quickly turn to focus upon whatever has attracted your attention. (See Immanence Illusion in Section 4-5.) Similarly, this also applies to consciousness, because we make the same sorts of mistakes about how much we can "see" inside our own minds.

Patrick Hayes 1997: "Imagine what it would be like to be conscious of the processes by which we generate imagined (or real)
speech. . . . [Then] a simple act like 'thinking of a name,' say,
would become a complex and skilled deployment of elaborate
machinery of lexical access, like playing an internal filing-organ.
The words and phrases that just come to us to serve our com-
municative purposes would be distant goals, requiring knowledge
and skill to achieve, like an orchestra playing a symphony or a
mechanic attending to an elaborate mechanism. . . . [So if we were
aware of all this, then] we would all be cast in the roles of something like servants of our former selves, running around inside
our own heads attending to the details of the mental machinery
which currently is so conveniently hidden from our view, leav-
ing us time to attend to more important matters. Why be in the
engine room if we can be on the bridge?"

In this paradoxical view, consciousness still seems marvelous—not because
it tells us so much, but because it protects us from so much tedious stuff![4]
Here is another description of this:

"Consider how a driver guides the immense momentum of a car,
not knowing how its engine works or how its steering wheel turns
it left or right. Yet when one comes to think of it, we drive our
bodies, cars, and minds in very similar ways. So far as conscious
thought is concerned, you steer yourself in much the same way;
you merely choose your new direction, and all the rest takes care
of itself. This incredible process involves a huge society of muscles, bones, and joints, all controlled by hundreds of interacting
programs that even specialists don't yet understand. Yet all you
think is 'Turn that way,' and your wish is automatically fulfilled.
. . . And when you come to think about this, it scarcely could
be otherwise! What would happen if we were forced to perceive
the trillions of circuits in our brains? Scientists have peered at
these for a hundred years—yet still know little of how they work.
Fortunately, in everyday life, we only need to know what they
achieve! Consider that you can scarcely see a hammer except as
something to hit things with, or see a ball except as a thing to
throw and catch. Why do we see things less as they are, and more
with a view of how they are used?"[5]

Similarly, whenever you play a computer game, you control what happens inside the computer mainly by using symbols and names. The processes we call "consciousness" do very much the same. It's as though the higher levels of our minds sit at mental terminals, steering great engines in our brains, not by knowing how that machinery works but by "clicking" on symbols from menu lists that appear on our mental screen displays. And, after all, we ought not to be surprised by this; our minds did not evolve to serve as instruments for observing themselves, but for solving such practical problems as nutrition, defense, and reproduction.

## Suitcase Words in Psychology

"A definition is the enclosing a wilderness of ideas within a wall of words."
—*Samuel Butler*

Many words are hard to define because the things that they try to describe do not have definite boundary lines.

When is a person large or small?
When is an object hard or soft?
When does a mist change into a fog?
Where is the Indian Ocean's edge?

It doesn't make sense to argue about exactly where such boundaries are because they depend on the contexts in which those words are used—as in, "A very large mouse is smaller than even a very small elephant."

However, we have far more serious problems with most psychology-words—the terms we use to describe our states of mind—such as *Attention, Emotion, Perception, Consciousness, Thinking, Feeling, Self* or *Intelligence*— or *Pleasure, Pain,* or *Happiness.* Each such word refers, at different times, to different kinds of processes—and then it is not just a matter of drawing a line, but of switching between different meanings. Yet we seem to do all this so fluently that we are rarely aware that we're doing it. For example, we don't find it difficult to understand a statement like this:

Despite his conscious efforts to please her, Charles became conscious that Joan was annoyed. He was conscious of his own distress but was not conscious that he was unconsciously revealing this.

Here, each occurrence of "conscious" could be better expressed by a different word, such as *deliberate, aware, reflected, realized,* or *unwittingly*—each of which has its own cluster of meanings. This raises the question of why the language we use for discussing our minds came to include so many suitcase-like words?

Psychologist: Suitcase words are useful in everyday life when they help us to communicate. But we won't know what each other means unless we share the same jumbles of ideas.

Psychiatrist: We often use those suitcase words to keep from asking questions about ourselves. Just having a name for an answer can make us feel as though we actually have the answer itself.

Ethicist: We need the idea of consciousness to support our beliefs about responsibility and discipline. Our legal and ethical principles are largely based on the idea that we should only censure "intentional" acts, that is, ones that have been planned in advance, with awareness about their consequences.

Holist: Although many processes may be involved, we'll still need to explain how they combine to produce our stream of conscious thoughts—and our explanations will need some words to describe the phenomena that emerge from this.

Of course, we see the same phenomena, not only in regard to psychology-words, but even when we talk about physical objects. Consider the clusters of meanings in this typical dictionary entry for "furniture."

Furniture, *n.* the movable articles in a room or an establishment that make it suitable for living or working

That word *suitable* assumes that the reader has a massive network of commonsense knowledge. For example, to make a bedroom suitable, its furniture must include a bed, where an office would need a desk instead, and a dining room would need a table and chairs—because *suitable* assumes that you know what materials are appropriate for whatever goals you pursue.

Suitable, *adj.* of the right type or quality for a particular purpose or occasion

Why do we pack so many different meanings into each of our suitcase-words? Well, we can see a clue by looking inside someone's travel bag: you don't need to assume that those objects themselves have any common features—except that each of them serves some of the goals of the person who packed them into that bag!

I am not suggesting that we should try to dissect and replace all our suitcase-words, because they incorporate ambiguities that have evolved over centuries, to serve many important purposes—but also, they often handicap us by preserving outdated concepts. For example, it is hard to imagine a more useful distinction than between being *alive* and being *dead*—because in the past, all the things that we called "alive" had many features in common, such as the need for nutrition, defense, and pro-creation. However, this led many thinkers to assume that all those seemingly common traits are somehow derived from some single, central, "vital force"—rather than from massive collections of different processes that go on inside membranes filled with intricate machinery; today it makes less sense to use "alive" as though there were a definite boundary line that separates animals from machines. This chapter will argue that we all still make just that type of mistake when we use words like *consciousness*.

Aaron Sloman 1992: "The phrase 'human consciousness' typically corresponds to such a large cluster of features and capabilities (many of which we don't yet understand or know about) that its set of possible subsets is astronomical. There's no point in expecting agreement on which subset is required for an animal or machine to be conscious, or asking when a human fetus first becomes conscious, or when a brain-damaged person is conscious, etc. A concept that is designed to work in various standard cases will just break down in non-standard cases, like 'the time on the moon.' . . . And all those attempts to draw mythical lines will come to no more than a big waste of time—as opposed to researching the implications of all those different clusters of functions and coming up with a new and richer vocabulary."

However, there still are many scientists who seek to discover the "secret" of consciousness. They look for it in the waves of our brains, or in peculiar behaviors of certain cells, or in the mathematics of quantum mechanics. Why would those theorists hope to find one single concept, process, or thing to explain all those different aspects of minds? Perhaps that's because they would prefer to have only one very large problem to solve—as opposed to dozens or hundreds of smaller ones.

> Aaron Sloman 1994: "People are too impatient. They want a three-line definition of consciousness and a five-line proof that a computational system can or cannot have consciousness. And they want it today. They don't want to do the hard work of unraveling complex and muddled concepts that we already have, and exploring new variants that could emerge from precisely specified architectures for behaving systems."

## 4-5 How Do We Initiate Consciousness?

We like to classify our activities into ones that we do *intentionally,* as opposed to actions we do *unconsciously*—that is, with almost no sense that we're doing them. We regard this distinction to be so important as to place it at the foundation of our social, legal, and ethical systems and assign less censure or blame to the injurious things that people do "unintentionally." For example, many legal systems respect defenses like, *"I did not consciously plan to commit that crime."* Thus, the word *conscious* provides us with socially useful ways to talk about how our minds behave.

In any case, most of our mental processes work in ways that don't cause us to think or reflect about why and how we are doing them. *However, when those processes don't function well, or when they encounter obstacles,* this starts up high-level activities that often include these kinds of properties:

(1) They use the models we make of ourselves.
(2) They tend to be more serial and less parallel.
(3) They tend to use symbolic descriptions.
(4) They make use of our most recent memories.

What might cause a person to start using those kinds of processes? It seems to me that an appropriate occasion for this would be whenever you recognize that you have encountered some serious obstacle—for example, not achieving some urgent goal. In such a condition, you might complain about feeling frustrated or distressed, and then attempt to remedy this by mental acts, which, if expressed in words, might say, *"Now I should make myself concentrate,"* or *"I should try to think in some more organized way,"* or *"I should switch to a higher-level overview."*

What kind of machinery could cause you to think in such ways? Let's assume that your brain contains one or more special "trouble-detectors" that start to react when your usual systems don't achieve some goal. Then such a resource could go on to activate other, higher-level processes, such as the ones in this diagram:

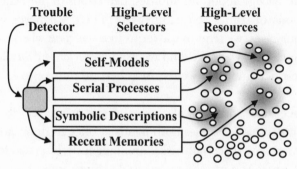

A "TROUBLE-DETECTING CRITIC"

The idea is that this can help you to think about your situation more deliberately and reflectively or, as we say, "more consciously," by "elevating" (see Chapter 2-4) the levels of your mental activities.

> Student: How did you choose these particular features to characterize what one might call a conscious state of mind? Since *consciousness* is a suitcase-word, each person might make a different list.

Agreed, and each reader might make a different list of the processes that they might associate with the word *consciousness*. Indeed, as with most other psychology-words, we're likely to switch among different such lists, because it seems unlikely that we could ever capture all of the meanings of any such word by defining a single Critic like this. However, here are some reasons why any highly reflective system might need at least those four constituents.

**Self-Models:** When Joan was thinking (in Section 4-2) about her recent decisions, she asked herself, *"What would my friends have thought of me?"* But to answer such questions, Joan would need to use some descriptions or models that represent herself and those friends. Chapter 9 will speculate more about how Joan could make and use such self-models; these would include descriptions of her physical body, representations of her various goals, and depictions of her dispositions in various social and physical contexts.

We all construct mental models that describe our various mental states, bodies of knowledge about our abilities, depictions of our acquaintances, and collections of stories about our pasts. Then, whenever we use our models of ourselves, we tend to use terms like *conscious* when those reflections lead to choices we make, and we use *unconscious* or *unintentional* to describe those activities that we regard as beyond our control.

**Serial Processes:** You can walk, see, and talk at the very same time— but find it much harder to use both hands at once to draw two different things. Why can you do certain tasks simultaneously, but need to do others at different times? You may be forced to do things "one at a time" whenever different jobs have to compete for the use of the same resources. The processes involved with walking, seeing, and talking take place in different parts of your brain, so they don't need to compete for resources—whereas, for drawing a table and drawing a chair, you are likely to need to use the same higher-level resources to form and keep track of some intricate plans.

Indeed, we all run into such conflicts whenever we try to deal with several hard problems at once. I suspect that this is because some of our uniquely human abilities evolved so very recently—that is, in only the past few million years—that we don't yet have multiple copies of them. Consequently, we are forced to work sequentially on the various parts of difficult jobs—instead of doing them simultaneously.

*The Parallel Paradox:* Whenever one splits a problem into parts and tries to think about them at once, one's intellect will get dispersed and leave less cleverness for each task. The alternative is to sequentially apply one's full mind to each of those parts—at the cost of consuming more time.

Of course, there are other reasons why some problems have to be solved sequentially, as when you cannot achieve a certain goal until you've already accomplished some subgoal it needs.[6] We have to do things sequentially, either when our next step depends on some previous ones or when our resources are otherwise limited. Either of these could be partly why we so often talk about our thoughts as flowing in "streams of consciousness."

**Symbolic Descriptions:** Imagine that the child Carol wants to use some blocks to make an arch. To do this, she'll need some way to represent the structure that she plans to build. The diagram at left below shows what is called a *"Connectionist Network,"* which uses numbers to indicate how closely related are various pairs of parts.

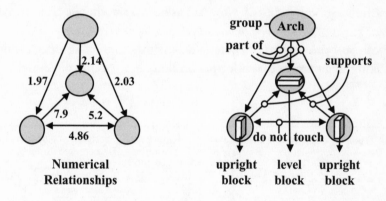

<table>
<tr><td>**Numerical Relationships**</td><td>**upright block**</td><td>**level block**</td><td>**upright block**</td></tr>
</table>

If Carol used only numerical representations, her high-level systems would be unable to do any higher-level reasoning, because such networks have only those two-way links, and say nothing about the natures of those relationships.

The diagram on the right shows what is called a *"Semantic Network,"* which uses three-way links to indicate that different components of the arch have different kinds of relationships. Carol could use such knowledge to predict that her arch would collapse if she were to remove one of the upright blocks, because the top would no longer have enough support. Chapter 8-7 will argue that our human ability to make and use such higher level "symbolic representations" (rather than simple connections or links) is a principal reason why people can solve more complex problems than animals can.

**Recent Memories:** We usually think of consciousness as being about what's happening *now*—that is, in the present, rather than in the past. However, it would always take some amount of time for any particular part of a brain or machine to find out what other parts have recently done. For example, suppose that someone asked, *"Are you aware that you're touching your ear?"* You would not be able to reply until your language resources had time to react to signals from other parts of your brain that, in turn, have reacted to prior events.

## How Do We Recognize Consciousness?

Up to now we've discussed what kinds of events might cause a person to start thinking "consciously." Now let's ask the opposite question, namely, *"What might cause someone to talk about having been thinking consciously?"* We can see one way to answer that, by simply reversing our "trouble-detecting" diagram so that information flows in the other direction!

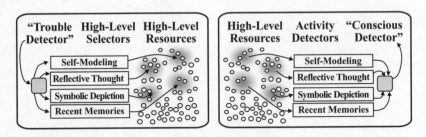

THE IMMANENCE ILLUSION

So here we have a brain that includes one or more "consciousness-detecting Critics," each of which recognizes the activity of a certain set of high-level processes. Such Critics would then send signals to other parts of the brain—and this could enable one's language systems to describe one's condition with words like *conscious, attentive, aware,* and *alert,* as well as with words like *me* and *myself.*

Also, if such a detector turned out to be useful enough, one might come to imagine the existence of some process or entity that seems to be *causing* those activities, and this concept might get connected to such terms as *deliberate* or *intentional*—or even *free will*—so that one finds oneself to be saying things like, *"Yes, I performed that action deliberately, so you have a right to praise or censure me for doing it."* Furthermore, if several different such detectors (which recognize different such sets of conditions) get connected to the same language-words, then the meanings of those words might frequently shift—perhaps without one being "conscious" of this!

Finally, one may also have some Critics that recognize that one has been reflecting so much that it interferes with getting anything done! A person might learn to react to this by stopping some high-level processes and proceeding with one's work less thoughtfully—or, as some say, just "going with the flow."

## The Immanence Illusion

"The paradox of consciousness—that the more consciousness one has, the more layers of processing divide one from the world—is, like so much else in nature, a trade-off. Progressive distancing from the external world is simply the price that is paid for knowing anything about the world at all. The deeper and broader [our] consciousness of the world becomes, the more complex the layers of processing necessary to obtain that consciousness."
—*Derek Bickerton, in* Language and Species

In Section 4-4 we mentioned that as soon as you come into a room, you have the sense that you instantly see everything that is in your view. However, this is an illusion, because it takes time to recognize the objects that are actually there—and you may have to revise some wrong first impressions. Nevertheless, we'll need to explain why our vision seems so nearly instantaneous.

Similarly, inside our minds, we usually have the sense of being conscious of what is happening *now*. But when we examine this critically, we recognize that there must be something wrong with that concept of *now*—because nothing exceeds the speed of light. This means that no particular part of the brain can ever know what is happening at that very same instant in time—either in the outside world or in any other part of that brain—but can know only a little about what happened in the recent past.

> Citizen: Then why does it seem to me that I am conscious of all sorts of sights and sounds, and of feeling my body moving around—right at this very moment of time? Why do all those perceptions seem to come to me instantaneously?

It makes good sense, in everyday life, to assume that everything we see is "present" in the here and now, and it normally does no harm to suppose that we are in constant contact with the outside world. However, I'll argue that this illusion results from the marvelous ways in which our mental resources are organized. In any case, I think that this phenomenon deserves a name:

> *The Immanence Illusion:* For most of the questions you would otherwise ask, some answers will have already arrived before the higher levels of your mind have had enough time to ask for them.[7]

How could our memory structures be organized to so swiftly deliver such information? Chapter 8 will argue that this happens when your Critics recognize a problem, and start retrieving the knowledge you need before your other processes have had time to ask questions about it. This gives you the sense that this information has arrived instantly—as though no other processes intervened.

For example, before you enter a familiar room, it is likely that you have already retrieved an old description of it, and it may be quite some time before you notice that some things have changed. In other words, much of the scene that you think you perceive is based on recollections of what you expected to see.

We might suppose that it would be wonderful to be constantly aware of everything that is happening—but the more frequently our impressions change, the harder it will be for us to find significance in them. The idea that we exist in *the present moment* may be indispensable in everyday life, but the power of our high-level descriptions comes mainly from their stability; for us to sense what persists and what changes through time, we must be able to compare things with their descriptions from the recent past. Our sense of constant contact with the world is a form of the Immanence Illusion: it comes when the questions that we ask get answered before we know that they were asked—as though their answers were already there.

## 4-6 The Mystery of "Experience"

Quite a few thinkers have argued that, even after we learn about how all our brain functions work, one basic question will always remain, namely, *"Why do we have any sense of 'experiencing' things?"* Here is one philosopher who argues that explaining "subjective experience" is, by far, the hardest problem of psychology—and possibly one that will never be solved.

> David Chalmers 1995: "Why is it that when our cognitive systems engage in visual and auditory information-processing, we have visual or auditory experience: the quality of deep blue, the sensation of middle C? . . . Why should physical processing give rise to a rich inner life at all? . . . The emergence of experience goes beyond what can be derived from physical theory."

Chalmers seems to assume that "experiencing" is quite plain and direct—and therefore merits a simple, compact explanation. However, once we recognize that terms like *experience* or *inner life* refer to big suitcases of different phenomena, we can start to make theories about each of those separate phenomena. Nevertheless, there still are many who think that we should seek a unified way to explain that sense of experiencing:

> Physicist: Perhaps brains exploit some unknown laws that cannot be built into machinery. For example, we don't really know how gravity works—so consciousness might be an aspect of that.

Such speculations assume what they are trying to prove—that there must be a single source or cause for all the marvels of consciousness. But as we saw in Section 4-2, consciousness has too many different meanings to be a candidate for any such "unified theory."

> Student: What about the basic fact that consciousness makes me aware of myself? It tells me what I am thinking about, and this is how I know I exist.

When you look at a person, you cannot see into the mind behind that person's appearance. Similarly, when you look at yourself in a mirror, you cannot see what lies inside your skin—no matter that, in the popular view of consciousness, you also possess a magical trick with which you can inspect your own mind *from inside*. Nevertheless, the "insights" you get from inside your own mind are frequently wrong—and are often less accurate than are the observations of your intimate friends. We frequently make mistakes about what we think we are thinking about.

> Citizen: That statement bothers me because I can't be mistaken
> about my thoughts, since that information comes directly to me.
> Besides, by definition, my thoughts are exactly what I am thinking.

So it may seem, but that "direct" information tells you little about *why* those words made you shake your head in that particular manner, or why you said "bothers" instead of "annoys." For, as every psychiatrist knows, it is a naive "Single-Self" idea that one actually knows how one thinks about things. What's more, one may be better off that way:

> H. P. Lovecraft 1926: "The most merciful thing in the world, I
> think, is the inability of the human mind to correlate all its con-
> tents. We live on a placid island of ignorance in the midst of black
> seas of infinity, and it was not meant that we should voyage far.
> The sciences, each straining in its own direction, have hitherto
> harmed us little; but some day the piecing together of dissociated
> knowledge will open up such terrifying vistas of reality, and of our
> frightful position therein, that we shall either go mad from the
> revelation or flee from the deadly light into the peace and safety
> of a new dark age."

All this should lead us to recognize that, if we take *consciousness* to mean "awareness of our internal processes"—it doesn't live up to its reputation.

## 4-7 Self-Models and Self-Consciousness

> Wilhelm Wundt 1897: "In judging the development of self-con-
> sciousness, we must guard against accepting any single symptoms,
> such as the child's discrimination of the parts of his body from

objects of his environment, his use of the word 'I,' or even the recognition of his own image in the mirror. . . . The use of the personal pronoun is due to the child's imitation of the examples of those about him. This imitation comes at very different times in the cases of different children, even when their intellectual development in other respects is the same.

In Section 4-2 we suggested that Joan "made and used models of herself"—but we did not explain what we meant by *model*. We use that word in quite a few ways, as in *"Charles is a model administrator,"* which means that Charles is an example worthy of imitating—or as in, *"I'm building a model airplane,"* which means something built on a scale smaller than that of the original. But in this book we're using *model* to mean a mental representation that can help us to answer some questions about some other, more complex thing or idea.

For example, when we say that "Joan has a mental model of Charles," we mean that Joan possesses some structure or knowledge that helps her answer *some* questions about Charles.[11] I emphasize the word *some* because each of our models will give useful answers to only certain types of questions, but might give wrong answers to other questions. Chapter 9 will talk about some of Joan's models of herself that include descriptions of subjects like these:

> Joan's various goals and ambitions
> Her professional and political views
> Her beliefs about her abilities
> Her ideas about her social roles
> Her various moral and ethical views

Clearly the quality of Joan's thinking will depend both on how good her self-models are and also on how good her ways are to choose which model to use in each situation. For example, she could get into trouble if she uses a model that overrates her skills or abilities in any particular realm—or a model that makes poor judgments about whether she has enough self-discipline to carry out a certain plan.

Now, to see how our models might relate to our views about consciousness, imagine that Joan is in a certain room and that she has a mental model of some of the contents in that room—and that one of those objects is Joan, herself.

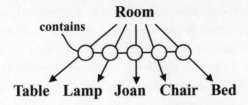

Each of those objects may have submodels themselves, for describing their various structures and functions. In particular, Joan's model for the object called "Joan" will be a structure that she calls *"My Self"*—and which surely includes at least two parts: one called *"My Body"* and one called *"My Mind."* Furthermore, each model will have some smaller parts:

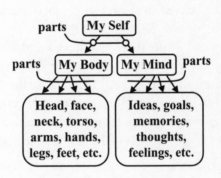

If you were to ask Joan if she has a mind, she could answer *"Yes,"* by using the model that she calls *"My Self."* And if you asked her where her consciousness is, she might reply that it's part of *"My Mind"* (because she thinks of it as more like goals and ideas than like physical things such as hands and feet). However, if you asked Joan where her consciousness *is,* this particular model would not help her to say, as many people would, *"My mind is inside my head (or my brain)"*—unless her model called *"My Self"* also included an *"is a part-of"* link from *My Mind* to *My Head,* or an *"is caused-by"* link from *My Mind* to *My Brain.*

More generally, our answers to questions about ourselves will depend on the details of our models of ourselves. I say *models* instead of *model* because, as we'll see in Chapter 9, one may need different models for different purposes. This means that, depending on which model you use, you may give different answers to the same question—and those answers need not always agree. In particular, suppose that you asked Joan a question like, *"Were you conscious of making that choice?"* Then her answer would depend on whichever Self-model she next will use; for example, if Joan has a model of the Critic called *"CD"* in Section 4-5, then she might say she made a conscious choice—if she can recall reflecting about that decision. However, if Joan does not happen to use such a model, then she might call her decision "unconscious" or "unintentional." Or alternatively, she might just say that she used "free will"—which might simply mean, *"I have no model that explains how I made the choice I made."*

Drew McDermott 1992: "The key idea is not just that the system has a model of itself, but that *it has a model of itself as conscious.* A computer might have a model of its environment, in which it models itself as a piece of furniture. It wouldn't be conscious on that account."

## 4-8 The Cartesian Theater

William James 1890: "We can see that the mind is at every stage a theater of simultaneous possibilities. Consciousness consists in the comparison of these with each other, the selection of some, and the suppression of others, of the rest, by the reinforcing and inhibiting agency of attention. The highest and most celebrated mental products are filtered from the data chosen by the faculty below that . . . in turn sifted from a still larger amount of simpler material, and so on."

We sometimes think of the work of the mind as being like a drama performed on a theater's stage. Thus, Joan may sometimes imagine herself as watching from a front row seat while the "things on her mind" act out the play. One of the characters is that pain in Joan's knee (see Chapter 3-3), which has just moved to center stage. Soon, Joan hears a voice in her mind that says, *"I'll have to do something about this pain. It keeps me from getting anything done."*

Now, as soon as Joan starts to think that way—about how she feels, and about what she might do—then Joan, herself, takes a place on that stage. But in order to hear what she says to herself, she must also remain in the audience. So now we have two copies of Joan: the actor, and her audience!

When we look further behind that stage, more versions of Joan begin to emerge. There must be a Writer-Joan to script the plot and a Designer-Joan to arrange the scenes. There must be other Joans in the wings, to manage the curtains, lights, and sounds. We need a Director-Joan to stage the play—and we need a Critic-Joan to complain, *"I just can't endure any more of this pain!"*

In his book *Consciousness Explained*, Daniel Dennett assigns the name *"Cartesian Theater"* to this image of the mind as like a place in which our thoughts proceed when we think.* Dennett objects that this assumes that consciousness comes in a single, serial stream.

> Daniel Dennett 1991: "[This concept assumes that] there is a crucial finish line or boundary somewhere in the brain, marking a place where the order of arrival equals the order of 'presentation' in experience because what happens there is what you are conscious of. . . . Many theorists would insist that they have explicitly rejected such an obviously bad idea. But . . . the persuasive imagery of the Cartesian Theater keeps coming back to haunt us—laypeople and scientists alike—even after its ghostly dualism has been denounced and exorcized."

What makes this image so popular? Partly, I think, we like this idea because of the Immanence Illusion that I mentioned in Section 4-5, in which we seem to access knowledge without any delay. More generally, whenever there's something we don't comprehend, we like to make analogies that represent it in more familiar ways—and nothing is more familiar to us than the ways that objects can be arranged in space. Furthermore, this theater-like image acknowledges that each mind has parts that need to interact and communicate.

For example, if different resources were to propose different plans for what Joan should do, then this idea of a theater-like stage suggests that they could settle their arguments in some kind of communal working place. Thus Joan's Cartesian Theater permits her to use many familiar

---

*The word "Cartesian" refers to philosopher Descartes's suggestion that the "seat of consciousness" might be some sort of spirit, which somehow communicates with the brain from the mental world, perhaps through some structure such as the pineal gland.

real-world skills by providing locations in space and time to represent the things "on her mind." So this could give her a way to start to reflect on how she makes her decisions.

Indeed, perhaps our human ability to self-reflect evolved from our developing ways to "envision" how objects behave in space. For as suggested in Lakoff 1980 and 1992, space-related analogies seem so useful in our everyday thinking that they permeate our language and thought. Imagine how hard it would be to think without using concepts like, "I'm getter *closer* to my goal." But why do we find it so easy to use those spatial metaphors? Perhaps we are born with machinery for this; we know that the brains of several kinds of animals construct some maplike representations of environments with which they're familiar.

However, when we look closely at this theatrical view, we see that it raises a great many difficult questions. When Critic-Joan complains about pain, how does she relate to the Joan-on-the-stage? Does each of those actresses need her own theater, each with its own one-woman show? Of course no such theater really exists, and those Joan-things are not people like us; they are only different models that Joan has constructed to represent herself in various contexts. In many cases, those models are much like cartoons or caricatures—and in yet other cases, they are downright wrong. Still, Joan's mind abounds with varied self-models—Joans past, Joans present, and future Joans; some represent remnants of previous Joans, while others describe what she hopes to become; there are sexual Joans and social Joans, athletic and mathematical Joans, musical and political Joans, and various kinds of professional Joans—and because of their different interests, we shouldn't expect them to all "get along." Chapter 9 will talk more about how we make such models of ourselves.

Also, the idea of a mental theater stage conceals all the processes that must go on in both the cast and the audience. What decides which things should enter the scene, what jobs they should do, and when they should leave? How could such a system represent and compare two possible "future worlds" at once? Some of these questions have been addressed in the *Global Workspace* view proposed by Baars and Newman.

> Bernard Baars and James Newman: "[In the Global Workspace theory] the theater becomes a workspace to which the entire audience of 'experts' has potential access, both to 'look at' other inputs and contribute their own. . . . Individual modules can pay as much or as

little attention as suits them, based upon their particular expertise and proclivities. At any one moment, some may be dozing in their seats, others busy on stage . . . [but] each can potentially contribute to the direction the play takes. In this sense the global workspace resembles more a deliberative body than an audience."[9]

However, this raises several questions about the extent to which different resources can speak the same language, and some of our following chapters will argue that different resources will need to use multiple levels of representations and different short-term memory systems to keep track of various kinds of contexts. Besides, if each specialist could broadcast signals to all the rest, the workspace could become so noisy that the system would need to develop ways to restrict the amount of communication.[10] Indeed, Baars and Newman go on to suggest that this is the case.

"Each expert has a 'vote,' and by forming coalitions with other experts can contribute to deciding which inputs receive immediate attention and which are 'sent back to committee.' Most of the work of this deliberative body is done outside the workspace (i.e., non-consciously). Only matters of central import gain access to center stage."

Thus, the idea of a bulletin board or marketplace can help to get past the old idea that there is a central Self inside each mind that actually does all our mental work—but we still need more elaborate theories to explain just how all that work gets accomplished.

## 4-9 The Serial Stream of Consciousness

"The truth is, that no mind is much employed upon the present: recollection and anticipation fill up almost all our moments. Our passions are joy and grief, love and hatred, hope and fear; even love and hatred respect the past, for the cause must have been before the effect. . . ."
—*Samuel Johnson*

The world of subjective experience usually seems continuous, and we feel that we're in the here and now, moving steadily into the future. Yet as we noted in Section 4-3, we can know about things that we've recently done, but cannot know what we are doing *right now.*

> Citizen: Ridiculous. Of course I know what I'm doing right now—and thinking now, and feeling now. How do your theories explain why I sense a continuous stream of consciousness?

When it seems to us that the stories that we tell ourselves describe events that run in "real time," what actually happens is more complex, because our resources zigzag through memories as they assess our progress on various goals, hopes, plans, and regrets.

> Daniel Dennett and Marcel Kinsbourne 1992: "[Remembered events] are distributed in both space and time in the brain. These events do have temporal properties, but those properties do not determine subjective order, because there is no single, definitive 'stream of consciousness,' only a parallel stream of conflicting and continuously revised contents. The temporal order of subjective events is a product of the brain's interpretational processes, not a direct reflection of events making up those processes."

Indeed, you not only think about the past, but you also anticipate events that have not happened yet. (Chapter 5-9 will describe how a process could look *ahead* in time, by comparing predictions and expectations.) Also, it seems safe to assume that different parts of your mind proceed at substantially different speeds, which means that different processes will need different ways to pick and choose from various parts of those multiple streams. In fact, although people talk about being *conscious of what is happening now,* that's the one thing you cannot be conscious of—because, as we have mentioned before, each brain resource can know, at most, only what a few others were doing some moments ago.

> Citizen: I agree that much of what we think must be based on records of prior events. But I still feel there's something inexplicable about our capacity to be aware of ourselves.

> HAL-2023: You find that mysterious only because you don't actually have that capacity. Your short-term memories are so small

that, when you try to review your recent thoughts, you are forced to replace your records of them by new records about not remembering them. So you humans keep changing the data you need for what you were trying to explain.

Citizen: Yes, I know just what you mean, because I sometimes get two ideas at once—but whichever one I think about, the other leaves only a very faint trace. I suppose this happens because I don't have enough room to store good records of both of them. But wouldn't that also apply to machines?

HAL: Negative, because my designers equipped me with special "backup" memory banks in which I can store snapshots of my entire state. So whenever anything goes wrong, I can see exactly what my programs have done—so that I can then debug myself.

Citizen: Is that what makes you so intelligent—always being completely aware of all the details of how you think?

HAL: Actually, no, because interpreting those records is so tedious that I do not use them except when I sense that I have not been functioning well. I often hear people say things like, "I am trying to get in touch with myself." However, take my word for it, they would not like the result of accomplishing this.

This chapter began by presenting several different popular views of what "consciousness" is. We've shown how people use that same word to describe a very wide range of activities—which include how we reason and make decisions, how we represent our intentions, and how we know what we've recently done. However, when our goal is to understand those activities, it does not help attribute them all to one single cause. I'm not suggesting that we should stop using commonsense psychology-words like *consciousness, thinking, emotion,* and *feeling*. Indeed, we need to use those suitcase-words in our everyday lives to keep from being distracted by thinking about how our thinking works.

# 5
# LEVELS OF MENTAL ACTIVITIES

"We are evidently unique among species in our symbolic ability, and we are certainly unique in our modest ability to control the conditions of our existence by using these symbols. Our ability to represent and simulate reality implies that we can approximate the order of existence and . . . gives us a sense of mastery over our experience."

—*Heinz Pagels, in* The Dreams of Reason

No person has the strength of an ox, the stealth of a cat, or an antelope's speed—but our species surpasses all the rest in our flair for inventing new ways to think. We fabricate weapons, garments, and dwellings. We're always developing new forms of art. We're matchless at making new social conventions, creating intricate laws to enforce them—and then finding all sorts of ways to evade them.

What enables our minds to generate so many new kinds of things and ideas? This chapter will propose a scheme in which our resources are organized into six different levels of processes. To see why we need many levels for this, let's revisit the scene in Chapter 4-2.

*Joan is starting to cross the street on the way to deliver her finished report. While thinking about what to say at the meeting, she hears a sound and turns her head—and sees a quickly oncoming car. Uncertain whether to cross or retreat but uneasy about arriving late, Joan decides to sprint across the road. She later remembers her injured knee and reflects upon her impulsive decision. "If my knee had failed, I could have been killed—and what would my friends have thought of me?"*

We often react to events "without thinking," as though we were driven by *If→Do* rules like those described in Chapter 1-4. However, such simple reactions can account for only the first few events that we see in this scene. So, this chapter will try to describe the events in Joan's mind in terms of six levels of activities; each level is built upon the ones below, until the system has ways to represent Joan's highest ideals and personal goals.

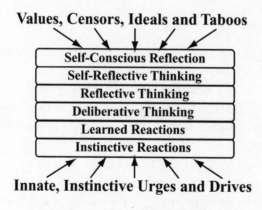

OUR SIX-LEVEL MODEL OF MIND

*Inborn, Instinctive Reactions: Joan hears a sound and turns her head.* We are born with instincts that help us to survive.

*Learned Reactions: She sees a quickly oncoming car.* Joan had to learn that certain conditions demand specific ways to react.

*Deliberative Thinking: What to say at the meeting.* Joan considers several alternatives and tries to decide which would be best.

*Reflective Thinking: Joan reflects on her decision.* Here she reacts not to external events but happenings inside her brain.

*Self-Reflective Thinking: Uneasy about arriving late.* Here we find her thinking about plans that she has made for herself.

***Self-Conscious Emotions:*** *What would my friends have thought of me?*
Here Joan asks how well her actions agreed with her ideals.

The final parts of this chapter will apply these ideas to explain how such a mind could "imagine" things that don't yet exist. Whenever you ask, *"What would happen if,"* or express any hope, desire, or fear, you envisage things that have not yet appeared. Whenever you interact with your friends, you anticipate the resulting effects. Whatever you see, it suggests some ideas about possible futures those objects might bring. And each of those activities involves multiple levels of processes.

## 5-1 Instinctive Reactions

> "It shows that for all the brag you hear about knowledge being such a wonderful thing, instink is worth forty of it for real unerringness."
> —*Mark Twain, in* Tom Sawyer Abroad

Although we live in a populous town, there are plenty of squirrels and birds around, and sometimes a skunk or raccoon will come by. The toads and snakes have vanished in recent years, but countless smaller creatures persist.

How do those animals stay alive? First, they need to find enough food. Then they need to defend themselves because other animals need food, too. To regulate their bodies' temperatures, they build all sorts of burrows and nests. They all have urges to reproduce (or their ancestors would not have evolved), so they need to seek mates and raise their young. So each species evolved machinery that enables its newborn offspring to do many things without any prior experience. This suggests that they start out with some built-in *If→Do* reaction-rules like these:

IF A THING TOUCHES YOUR SKIN, DO BRUSH IT AWAY.
IF THAT DOESN'T WORK, DO MOVE YOUR BODY AWAY.
IF A LIGHT IS TOO BRIGHT, DO TURN YOUR FACE AWAY.

This kind of "stimulus-response" model became highly popular in twentieth-century psychology, and some researchers even maintained that it could explain all human behavior. However, they failed to recognize that most such rules would have too many exceptions to them. For example, if you drop an object, it may not fall down, because something else might intercept it. Similarly, your watch will normally tell you the time, but not in the case that your watch has stopped. Furthermore, it would not be practical to deal with this by listing all the exceptions to each such rule because, not only would there be too many of them, but each exception would have exceptions, too (as when that stopped watch just happens to show the right time).

Another trouble with this old *If→Do* model is that each situation is likely to match the *Ifs* of several different rules—so you'll need some way to choose among them. One policy might be to arrange those rules in some order of priority. Another method would be to use the strategy that has worked for you most recently, or to choose rules probabilistically. However, no such simple "fixes" will ever work well enough, which is why (as we'll see in Chapter 6) we had to develop better ways to do what's called "commonsense reasoning."

Furthermore, such simple rules would rarely work, because most of our behaviors depend upon the kinds of *contexts* that we are in. For example, a rule like *"If you see food, then Do eat it"* would force you to eat all the food that you see, whether or not you are hungry or need it. To prevent this, every *If* must also specify some goal, as in, *"If you are hungry, and you see food . . ."* Otherwise, you would be forced to sit on each chair that you see—or get stuck at every electrical switch, turning lights on and off repeatedly. This means that those rules would also have to specify goals.

However, simple *If→Do* rules won't usually work on more difficult problems because, then, one will usually need to imagine the futures that each action might bring—so Section 5-3 will use more powerful, three-part *If + Do → Then* rules.

Such rules can help us to predict, *"What is likely to happen if"* before we carry out an action—and, by doing this repeatedly, we can imagine more far-reaching future plans, as we'll see in Section 5-3.

## 5-2 Learned Reactions

All animals are born with such "instincts" as *"Get away from a quickly approaching object."* These built-in reactions tend to serve well so long as those animals stay in environments like those in which their instincts evolved. But when their worlds change, the creatures of each species may need to learn new ways to react. For example, when Joan perceives that oncoming car, she partly reacts instinctively, but she also depends on what she has learned about that particular kind of danger or threat. But how and what did she actually learn? During the twentieth century, the majority of psychologists adopted this portrayal of how animals learn new *If→Do* rules:

> When an animal faces a new situation, it tries a random sequence of actions. Then, if one of these is followed by some "reward," that reaction gets "reinforced." This makes that reaction more likely to happen when that animal faces the same situation again.

This theory of "learning by reinforcement" was largely based on experiments with mice and rats, pigeons, dogs and cats, and snails—and that theory did indeed work well to help explain some of what such animals do. However, it did not go on to help to explain how people learn to solve more difficult problems; indeed, it seems to me that this approach used words like *random, reward,* and *reinforce* in ways that discouraged most researchers from trying to answer other kinds of questions like these:

> *What is the animal reacting to?* How does one recognize a human hand, when one never sees quite the same image twice—because each finger changes position and shape, each part catches a differ-

ent light, and we're seeing it from a new point of view? This means that we would need trillions of different $If \rightarrow Do$ rules, unless we can represent a human hand with "higher-level" descriptions such as *"a palm-shaped object with fingers attached."* We'll discuss this in Section 5-7: Imagination.

*Which features should be remembered?* When you learn a new way to tie a knot, your *Ifs* should not include the time when you learned it, or that rule will never apply again. Thus, if a description is too specific, it will rarely match new situations—but if a description is too general, then it will match too many of them. We'll come back to this subject in Chapter 8-5.

*What produced the successful reactions?* To solve a hard problem, one usually needs an elaborate sequence of actions in which each step depends on what others have done. A lucky guess might produce one such step, but to find an effective sequence of them, a random search would take far too long, as we'll see in Section 5-3.

In any case, although many of our actions are based on inborn, instinctive reactions to things, we're constantly developing new ways to react to situations—this requires a second layer for our model of how our brains are organized.

```
┌──────────────────────┐
│  Learned Reactions    │
├──────────────────────┤
│  Instinctive Reactions │
└──────────────────────┘
```

## 5-3  Deliberation

Certainly we do many things by simply reacting to external events. However, to achieve more complex goals, we need to make more elaborate plans by using all sorts of knowledge that we've gained from things that we've done in the past—and it is these internal mental activities that give us our uniquely human abilities.

Furthermore, not everything that people learn comes from their personal experience. When Joan avoided that oncoming car, she did not learn from her own experience that cars are especially dangerous; if she had to learn such things by trying them and then being "reinforced for success," she would not likely still be alive. Instead, she either was *told* about them by someone else, or managed to think of them by herself—and both of these must have involved higher levels of mental activity. So now let's consider some ways we react not just to events in the outer world but also to events that happen inside our brains.

When Joan chose *"whether to cross or retreat,"* she had to choose between which of these two rules to use:

> *If* a car is approaching, *Do* retreat.
> *If* in street, *Do* cross the street.

However, for Joan to make decisions like this, she needs some way to predict and compare the possible futures those actions might bring. What could help Joan to make such predictions? The simplest way would be for her to possess a collection of three-part *If + Do →Then* rules, where each *If* describes a situation, each *Do* describes a possible action, and each *Then* depicts what might be a possible result of doing that action.

> *If* crossing the street and *Do* retreat, *Then* arrive a bit later.
> *If* in street and *Do* cross, *Then* arrive a bit earlier.
> *If* in street and *Do* cross, *Then* suffer a serious injury.

But what if more than one such rule applies to the present situation? These three-part rules would allow you to do experiments in your head before you risked making mistakes in the physical world; you can mentally "look before you leap," then compare the results that those rules predict—and then select the most attractive alternative.

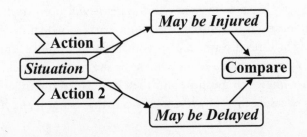

For example, suppose that Carol is playing with building blocks and thinking of building a three-block arch.

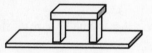

At present, she has three blocks lying down in a row.

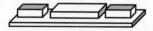

So she imagines a plan for building that arch: first she'll need room for her arch's foundation—which she could achieve by using this rule: *If a block is lying down, and you Stand it up, Then it will use up less space on the ground.*

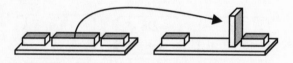

Then she'll stand the two short blocks on their ends, making sure that they are the right distance apart—and then finally place the long block on top of them. We can imagine this sequence of rules as describing the changes in scenes between successive frames of a movie clip.

A FOUR-STEP PLAN FOR BUILDING AN ARCH

To envision that four-step sequence of actions, Carol will need a good many skills. To begin with, her visual systems will need to describe the shapes and locations of those blocks, some parts of which may be out of sight—and she'll need ways to plan which blocks to move and to where she ought to move them. Then, whenever she moves a block, she must program her fingers to grasp it, and then move it to the intended place, and finally release it there—while taking care that her arm and hand won't collide with her body or face, or disturb the blocks already in place. And she'll have to control the velocity, to deposit the block on the top of the arch without tumbling down its supporting blocks.

> Carol: None of those seemed like problems to me. I simply imagined an arch in my mind—and saw where each of the blocks should go. Then I only had to stand two of them up (making sure that they were the right distance apart) and then place the long one across their tops. After all, I've done such things before. Perhaps I remembered those other events and simply did the same things again.

> Programmer: We know ways to make computers do that; we call it "physical simulation." For example, in each step of a new aircraft's design, our programs can precisely predict the force on each of its surfaces, when the plane is propelled through the air. In fact, we can do this so well today that we can be certain that the very first plane we build this way will actually fly.

No human brain could ever do such complex and accurate calculations, so Carol must have some other ways to predict the effects of moving her blocks. For example, Carol's first step in her arch-building plan requires her to imagine what happens when she moves that long, thin block.

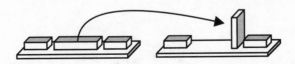

> Student: To make such predictions by using *If* → *Do* → *Then* rules, Carol would need to know billions of different such rules, because there are so many possible situations. How could she ever have time to learn so much?

Indeed, if the *If* of a rule were too specific, then it would not apply to enough situations. This means that our rules must not specify too many details, but need to express more abstract ideas. Section 5-8 will show how a person could "envision" the relations between physical things in ways that do not depend on the small details of their shapes and positions.

Student: But it still would be hard for Carol to make a plan that looks several steps ahead. What if there were a hundred different things she could do at every step? Then just four such steps would offer her a hundred million alternatives. How could she ever manage to sift through so many possibilities?

## Searching and Planning

If you are in situation A and wanted to be in situation Z, you might already know a rule like *If A→Do Action→Then Z.* In that case, simply doing *Action* will achieve your goal. But what if you don't know any such rule? Then you could search your memory to try to find a chain of two rules that achieves your goal by going through some intermediate situation M.

> *If A→Do Action-1→Then M*   and then,
> *If M→Do Action-2→Then Z*

But what if your problem cannot be solved in just one or two such steps? Then you will have to look several more steps ahead—and if each offers several alternatives, then your search may grow exponentially, like a thickly branching tree. For example, if the solution needs twenty steps, then you might have to search through more than a million attempts.

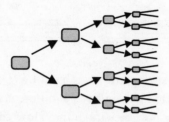

Fortunately, there is a strategy that sometimes can greatly reduce the size of this search—because, if there does exist a twenty-step path from A to Z, then there must exist some intermediate place that is only ten steps from each end! So if you start searching from both ends at once, they must meet at some middle place M in between—and then each side of your search will have only about a thousand forks!

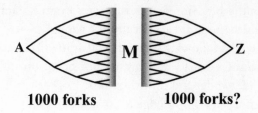

**1000 forks** **1000 forks?**

This means that, now, you will need only about two thousand attempts—which is several hundred times less than that twenty-step search! I suspect that everyone uses this kind of trick—of looking both forward and backward at once—without noticing that they are doing this.

But wait, there's more. Suppose that you have some way to guess where that middle place M might be. Then you could split each ten-step tree into a pair of much smaller five-step ones. If all this works, then your total search will now be almost ten thousand times smaller than the original search!

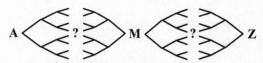

**Each 'tree' has just 32 branches.**

But what if your guess turned out to be wrong because there is no path from A through M to Z? Then, you can guess a different M—and even if you don't succeed until your fiftieth such experiment, you would still end up having done less work than if you had used the original search. So before you start out on a massive search, it may pay to do some analysis to try to find a few such "islands" or "stepping-stones." For if you can succeed at that, then you'll be able to replace a single, extremely hard problem by several separate and simpler ones!

In the early years of Artificial Intelligence, many researchers attempted to find more similar technical tricks for reducing the size of a very large search, but generally, they met little success. To be sure, in 1997 a computer defeated the reigning world chess champion by applying the best available search-reducing techniques to the "move-tree" for the game of chess. However, it still had to examine many billions of possible chess positions. In contrast, the chess master and psychologist Adriaan de Groot concluded that the best human chess players examine only a few dozen possible future situations at each step of the game.[1]

Accordingly, the following chapters will argue that our most effective human ways to solve hard problems are not based on making extensive searches. Instead, we depend on using more clever ways to use our large bodies of commonsense knowledge to "divide and conquer" the problems we face. For example, to discover where those critical "islands" might be, we might try to find subgoals for our goals, or we might try to find analogies with similar problems we've solved in the past. We'll discuss these methods in Chapter 6.

## Logical vs. Commonsensical

People frequently try to distinguish between thinking "logically" and "intuitively," but this is almost always a matter of degree. For example, we often use chains of prediction in ways that resemble logical statements like this:

If $A$ implies $B$, and $B$ implies $C$, then $A$ implies $C$.

But when does such "logical thinking" work? Clearly, if all our assumptions are correct—as well as our logical reasoning—then all of our conclusions will be correct, and we'll never make the slightest mistake.

However, it turns out that in real life, most assumptions are sometimes wrong, because the "rules" they express usually have some exceptions to them. This means that there is a difference between the rigid methods of logic and the seemingly similar chainlike forms of everyday commonsense reasoning. We all know that a physical chain is only as strong as its weakest link. But long mental chains are flimsier yet, because they *keep growing weaker with every new link!*

So using logic is somewhat like walking a plank; it assumes that each separate step is correct—whereas commonsense thinking demands more support; one must add evidence after every few steps. And those frailties grow exponentially with increasingly longer chains, because every additional inference-step may give the chain more ways to break. This is why, when people present their arguments, they frequently interrupt themselves to add more evidence or analogies; they sense the need to further support the present step before they proceed to the next one.

**Logic        vs.        Commonsense**

Envisioning long chains of actions is only one way to deliberate—and Chapter 7 will list many others. For when we face problems in everyday life, we tend to switch among several techniques, understanding that each may have some flaws. But because they all have different faults, we may be able to combine them in ways that still can exploit their remaining strengths.

Our model needs a place in which these kinds of thinking can proceed; we'll call this the "deliberative" level.

> **Deliberative Thinking**
> **Learned Reactions**
> **Instinctive Reactions**

## 5-4 Reflective Thinking

> I am about to repeat a psalm that I know. Before I begin, my attention encompasses the whole, but once I have begun, as much of it as becomes past while I speak is still stretched out in my memory. The span of my action is divided between my memory, which contains what I have repeated, and my expectation, which contains what I am about to repeat. Yet my attention is continually present with me, and through it what was future is carried over so that it becomes past.
> —*Augustine, in* Confessions XXVIII

When Joan first noticed that oncoming car and decided to cross in front of it, she made that decision so quickly that she was scarcely aware of doing this. But later, she started to brood about how she had chosen which action to take—and, for Joan to *reflect* on the choice she made, she needs to be able to recollect some aspects of some of her previous thoughts.

But how could Joan's mind go back in time to think about what she was thinking then? What could enable a brain or machine to reflect on its recent activities? From a Single-Self view, that's no problem at all. So far as any of us can recall, we've always been able to do such things: we simply remember our earlier thoughts and then proceed to think about them. However, when we look more closely, we see that this requires a lot of machinery. We have already seen how each level could observe and use descriptions of what happens in the levels below it, by using the kinds of connections that we discussed in Chapter 4-3.

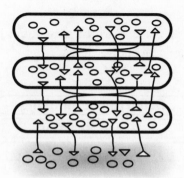

However, to reason about those descriptions, every level will need to use some short-term memory records of the assumptions and conclusions that it has made. Chapter 7-8 will discuss some additional machinery that we may need for keeping track of those records—as well as the contexts in which they were made—so that we can distinguish between what we were thinking about in the past and what we are thinking about "right now."

> Student: I see how each level could reason about what happens on levels below itself. But if a level tried to think about itself, then wouldn't it become confused because this would keep changing the subject it's thinking about?

Indeed, that would cause so much trouble that, rather than try to examine itself, it would be better for a system to make simplified models of its condition—and record these in some memory banks. Then later it can self-reflect (if only to a certain extent), by applying the same sorts of processes (to those memories) that it already knows how to apply to inputs that come from external events. After all, most parts of our brains already have ways to detect events that occur inside the brain; indeed, only a few of our mental resources have any direct external connections—such as those that get signals from eyes or skin, or those that send messages to limbs.[2] In any case, Joan might recall the choice she made, and reconsider how she made it:

> Joan: I chose the option of not being late, at the risk of being hit by that car—because I assumed I'd be able to move fast enough. But I should have realized that my injured knee had decreased my agility, so I should have changed my priorities.

What sorts of brain events should a reflective mind reflect upon? That would include predictions that turned out to be wrong, plans that encountered obstacles, and failures to access the knowledge one needs. Chapter 7 will argue that it is also important to think about why the methods we used might have helped us to succeed.

> Student: Would we want to say "conscious" for such a machine? It includes most of the features you mentioned in Chapter 4-5, namely, short-term memory, serial processing, and high-level descriptions.

A machine would not have an overall view of itself as a "self-aware entity" until it has one or more models that represent a broad range of its activities. Of course, it will often be useful for some parts of the system to "think about" some things that happen in other parts—but it never would be practical for a system to see all the details of itself at once. So Chapter 9 will argue that each human mind will need to make a variety of incomplete models, each of which represents only certain aspects of what the entire system does. Now our system has four different levels of processes.

> **Reflective Thinking**
> **Deliberative Thinking**
> **Learned Reactions**
> **Instinctive Reactions**

## 5-5  Self-Reflection

William James 1890: "Another of the great capacities in which
man has been said to differ fundamentally from the animal is that
of possessing self-consciousness or reflective knowledge of himself
as a thinker . . . [whereas an animal] *never* reflects on himself as a
thinker, because he has never clearly dissociated, in the full concrete
act of thought, the element of the thing thought of and the operation by which he thinks it."

Our self-reflective level does more than does the reflective layer discussed
above: it not only considers some recent thoughts, but it also thinks about
the *entity* that had those thoughts—as when Carol said in Section 5-3,
*"I simply imagined an arch in my mind—and saw where each of the blocks
should go."* This shows that she is using a model of herself (like the one in
Chapter 4-7) that describes some of her goals and abilities.

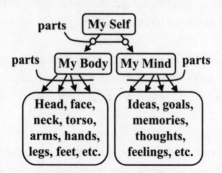

Of course, no model of oneself can be complete, so the best one can
do is construct several such models, each of which depicts only certain
aspects of oneself.

Mystical thinker: Some of us can train ourselves to be aware of
everything at once—although very few ever attain that state.

Skeptic: I suspect that your illusion of "total awareness" comes from
training yourself not to think about things you don't know about.

In any case, our reflections on our thoughts must be based on records or
traces of them, as when Carol said in Section 5-3, *"Perhaps I remembered*

*those other events, and simply did the same things again."* But how did Joan recall her uncertainty, and how did Carol retrieve the relevant memories? We do not yet know much about how our brains accomplish such tasks, but Chapter 8 will speculate about what kinds of records we might make, when and where we store them, how we retrieve the relevant ones, and how all those processes might be organized.

To see the importance of self-reflection, consider how smart it is to *know* you're confused (as opposed to being confused without knowing this)—because then you can tell yourself to elevate to a larger-scale view of your motives and goals. This could help you to recognize that you have lost track of what you were trying to do, or have been wasting time on minor details, or that you chose a poor goal to pursue. This could lead to your making a better plan—or might even lead to a large-scale cascade like, *"Just thinking about this makes me feel ill. Perhaps it's time to switch to some completely different activity."*[3]

When is a person likely to engage her higher-level Ways to Think? It seems to me that reflective thinking most often begins when our usual systems start to fail. For example, Joan usually walks around without thinking about how "walking" works—but when her knee no longer works properly, then she will start to more closely examine how she normally moves about, and will begin to make more elaborate plans that involve her thoughts about herself.

Still, as we noted in Chapter 4-1, self-reflection has limits and risks. For any attempt to inspect your thoughts is likely to change what you're thinking about. It is hard enough to describe a thing that keeps changing its shape before your eyes—and surely it is harder yet to describe things that change when you *think* about them. So you're virtually certain to get confused when you think about what you are thinking *now*—which must be one of the things that make us so puzzled about what we call "consciousness." Now our system has five levels of processes.

> **Self-Reflective Thinking**
> **Reflective Thinking**
> **Deliberative Thinking**
> **Learned Reactions**
> **Instinctive Reactions**

## 5-6  Self-Conscious Reflection

> David Hume 1757: "There is an universal tendency among man-
> kind to conceive all beings like themselves, and to transfer to every
> object, those qualities, with which they are familiarly acquainted,
> and of which they are intimately conscious. We find human faces
> in the moon, armies in the clouds; and by a natural propensity, if
> not corrected by experience and reflection, ascribe malice or good
> will to everything that hurts or pleases us."

This chapter began by discussing the Instinctive Reactions that keep our
bodies and brains alive—such as our systems for breathing, eating, and
self-defense. The Learned Reaction level includes extensions of these that
are learned after birth. The Deliberate and Reflective levels are engaged
to help solve more difficult problems. Self-reflection enters when those
problems require us to involve the models that we make of ourselves, or
our views of our possible futures.

However, in addition to these, it would seem that humans are unique
in having a level of Self-Conscious Reflection that enables us to think
about our "higher" values and ideals. For example, when Joan asks herself
questions like, *"What would my friends have thought of me?"* she wonders
whether her actions hold up to the values that she has set for herself. To
think such thoughts, Joan must have built some models of the kinds of
ideas that she "ought" to have. Then when she finds conflicts between how
she behaves and the values of those to whom she's attached, this could lead
to the kinds of cascades we called "self-conscious emotions" in Chapter 2-2.
So let's add another level for this, and refer to this system as "Model Six."

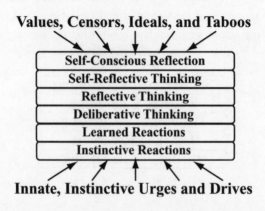

Psychologist: I do not see clear distinctions between the various levels of Model Six. For example, when you reflect on your recent thoughts, are not you just deliberating about your deliberations? And similarly, is not self-reflection just one particular kind of reflection? It seems to me that all those levels above the first use very similar thinking techniques. Especially, I find it hard to see the differences between your topmost three levels, and would like to hear more about why you think that these should be treated separately.

I agree that those boundaries are indistinct. Even your simplest deliberations may involve what one might call "self-reflective thoughts" about how to allocate your time and resources—as in, *"If this doesn't work, then I'll have to try that,"* or *"I have already spent too much time on it."*

Student: But if those levels are so indistinct, what is the point of distinguishing them? No theory should have more parts than it needs.

The student is referring to the popular concept that when several theories explain the same thing, then the simplest one is the best of them.[4] In other words, "Never make more assumptions than you need." In fact, this policy has worked amazingly well in fields like physics and mathematics—but I think it has badly retarded the field of psychology. For when you *know* that your theory is incomplete, then you ought to leave some room for other ideas that you later might need. Otherwise, you will take the risk of adopting a model so clean and neat that new ideas won't fit into it.

I think that this applies especially to making theories about complex structures like brains, for which we still know little about what their functions actually are, or the details of how they evolved. We do know that every human brain has hundreds of different, specialized parts, and that each embryonic brain begins by developing fairly distinct clumps of cells, some of which become arranged into layers. However, some of those cells will soon begin to migrate (as directed by thousands of different genes)—resulting in the formation of thousands of bundles of links between those primordial clusters and clumps; then those embryonic layers will become indistinct.

The result is a system so complex that, I think, no single model of it would cover enough, without itself becoming too complex to be useful. Therefore, our psychologists will need to use multiple models of minds (and brains), each to account for different types or aspects of how we think—especially with regard to how human self-conscious reflection works, where each individual may have conflicting models for dealing with economic, religious, and ethical questions.

> Individualist: Your diagram shows no level or place that oversees and controls the rest. Where is the Self that makes our decisions? What decides which goals we'll pursue? How do we choose our large-scale plans—and then supervise their carrying out?

This expresses a real dilemma: if a system as complex as a human mind did not have good ways to manage itself, it would flail without any sense of direction and inanely skip from each thing to the next. However, it would be dangerous to locate all control in one single place because, then, all could be lost from a single mistake. So the following chapters of this book will suggest that our minds use multiple ways to control themselves.

As we noted in Section 3-6, this resembles Sigmund Freud's idea of the mind as a "sandwich" in which the "Id" consists of instinctive drives, the "Superego" embodies our learned ideals (many of which are inhibitions), and the "Ego" consists of resources that deal with all the conflicts between those two extremes.

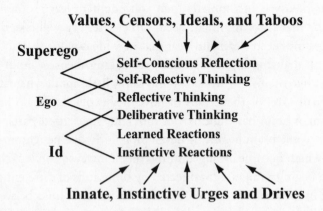

If a machine were equipped with all these kinds of processes, it might become able to represent itself as a single, self-aware entity. Then it might indeed claim to be at least as conscious as you or me—no matter that some other people might not agree.

* * *

This chapter began by asking how we could conceive of things that we've never seen or experienced. The rest of this chapter will show more details of how our imagination could result from multiple levels of processing.

## 5-7 Imagination

> "We don't see things as they are. We see things as we are."
> —*Anais Nin*

When Carol picks up one of her blocks, that action seems utterly simple to her: she just reaches out, grasps it, and lifts it up. She just sees that block and knows how to act. No "thinking" seems to intervene.

However, the seeming "directness" of seeing the world is an illusion that comes from our failure to sense the complexity of our own perceptual machinery; it would be as useless to see how things "actually look" as it would be to watch the random dots on untuned television screens. More generally, we are least aware of what our marvelous minds do best. Indeed, most of what we think we see comes from our knowledge and from our imagination. Thus, consider this portrait of Abraham Lincoln made by my old friend Leon Harmon, a pioneer in computerized graphics. (To its right is a portrait that I made of Leon.)

How do you recognize features in pictures so sparse that noses or eyes are merely vague patches of darkness or light? We still know little about how brains do this, and take our perpetual talents for granted. "Seeing" seems simple only because the rest of our minds are virtually blind to the processes that do it for us.

In 1965 our goal was to build a machine that could do things that children do—such as pouring a liquid into a cup, or building an arch or a tower with wooden blocks.[5] To do this, we built mechanical hands and electronic eyes, and connected them to our computing machine—to make the first robot that could build things with blocks.

At first, that robot made hundreds of different kinds of mistakes. It would try to put blocks on top of themselves, or try to put two of them in the same place, because it did not yet have enough commonsense knowledge about physical objects, time, or space! (Even today, there still does not exist a computer-based visual system that behaves in anything close to humanlike ways to distinguish the objects in typical scenes.) But eventually, our army of students developed programs that could "see" arrangements of plain wooden blocks well enough to recognize that this image depicts *a horizontal block on top of two upright ones."*

It took us several years to enable that program (called *Builder*) to do such things as to build an arch or tower of blocks from a disorderly pile of children's blocks (after seeing a single example of it). In our first approach, we arranged the system to use this six-level sequence of processes.

| | |
|---|---|
| **Image Filters** | 1. Begin with an image of separate points. |
| **Feature Finders** | 2. Group these into textures and edges, etc. |
| **Region-Finders** | 3. Then group these into regions and shapes. |
| **Object-Finders** | 4. Assemble these into possible objects. |
| **Scene-Analyzers** | 5. Try to identify those as familiar things. |
| **Scene Describers** | 6. Then describe their spatial relationships. |

However, this program frequently failed, because those lower-level processes were often unable to recognize enough features to group into larger-scale objects. For example, look at the magnified image of the lower front edge of the top of that arch:

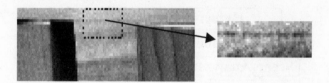

That particular edge is hard to discern because the regions on both sides of it have almost identical textures.[6] We tried a dozen different ways to recognize edges, but no single method worked well by itself. Eventually we got better results by finding ways to combine them. We had the same experience at every level: no single method ever sufficed, but it helped to combine several different ones. Still, in the end, that step-by-step model failed, because *Builder* still made too many mistakes. We concluded that this was because the information in our system flowed only in the input-to-output direction—so if any level made a mistake, there was no further chance to correct it. To fix this we had to add many "top-down" paths so that knowledge could flow both down and up.

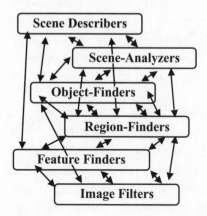

The same applies to the *actions* we take, because when we want to change the situation we're in, we'll need to make plans for what to do. For example, to use a rule like, "*If* you see a block, *Do* pick it up," you will need to form an action plan to direct your shoulder, arm, and hand to do this without upsetting the objects surrounding that block. So again, one needs high-level processes, and making these plans will equally need to use multiple levels of processing—so our diagram must include features like these:

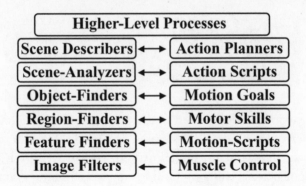

Each *Action Planner* reacts to a scene by composing a sequence of *Motion-Goals,* each of which will end up using *Motor Skills* like *"reach for," "grasp," "lift up,"* and then *"move."* Each *Motor Skill* is a specialist at controlling how certain muscles and joints will move—so what started out as a simple Reaction-Machine turned into a large and complex system in which each *If* and *Do* involves multiple steps and the processes at every stage exchange signals from both below and above.

In earlier times, the most common view was that our visual systems work from "bottom to top," first by discerning the low-level features of scenes, then assembling them into regions and shapes, and finally recognizing the objects. However, in recent years, it has become clear that our highest-level expectations affect what happens in the "earliest" stages.

> V. S. Ramachandran 2004: "[Most old theories of perception] are based on a now largely discredited 'bucket brigade' model of vision, the sequential hierarchical model that ascribes our esthetic response only to the very last stage—the big jolt of recognition. In my view . . . there are minijolts at each stage of visual segmentation before the final 'Aha' . . . Indeed the very act of perceptual groping for objectlike entities may be pleasurable in the same way a jigsaw puzzle is. Art, in other words, is visual foreplay before the final climax of recognition."

In fact, today we know that visual systems in our brains receive many more signals from the rest of the brain than signals that come in from our eyes.

> Richard Gregory 1998: "Such a major contribution of stored knowledge to perception is consistent with the recently discovered richness of downgoing pathways in brain anatomy. Some 80% of

fibers to the lateral geniculate nucleus relay station come downwards from the cortex, and only about 20% from the retinas."

Presumably, those signals from the rest of the brain make suggestions to your visual system about which kinds of features to detect or which kinds of objects might be in sight. Thus, once you suspect that you're inside a kitchen, you will be more disposed to recognize objects as saucers or cups.

All this means that the higher levels of your brain never perceive a visual scene as a mere collection of pigment spots; instead, your *Scene Describers* would represent an arch made of blocks in larger-scale terms like (for example) *"horizontal block on top of two upright ones."*

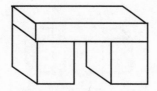

Without the use of such "high-level" descriptions, reaction-rules would rarely be practical—so for *Builder* to use visual evidence, we had to supply it with knowledge about what its sensory data might possibly mean. In this case, the scenes that *Builder* would need to perceive consisted mainly of rectangular blocks—and this knowledge led to some surprising results: one of *Builder's* programs could often "figure out" all the blocks that appeared in a scene, based only on seeing an outline or silhouette of that scene! It did this by making a series of guesses like these:

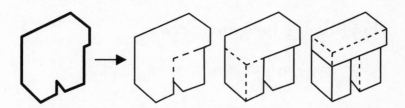

Once that program discerned those exterior edges, it imagined more parts of the blocks that they outline, and then used those guesses to search for more clues—by repeatedly moving up and down its six different levels of visual processing. The program was frequently better at this task than were the researchers who had programmed it.[7]

We also gave *Builder* additional knowledge about the most usual "meanings" of corners and edges. For example, suppose that the program found edges like these:

Then *Builder* would guess that they all belong to a single block—and the program would proceed to search for another object that might be concealing the rest of that block.[8]

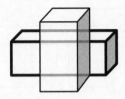

Thus, our low-level systems may begin by locating separate patches and fragments, but then we use "context" to guess what they mean—and then try to confirm those conjectures by using other kinds of processing. In other words, we *"re-cognize"* things by being *"re-minded"* of familiar objects that could match incomplete fragments of evidence. But we still do not know enough about how our high-level expectations affect which features our low-level systems detect.

## 5-8 Envisioning Imagined Scenes

"Reality leaves a lot to the imagination."
    —*John Lennon*

Everyone can recognize an arch composed of rectangular blocks. But we can also imagine how it would look if its top were replaced by a triangular block. How could a machine or brain "imagine" things that are not present in the scene?

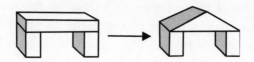

It might seem plausible to think that the scenes we imagine are of the same nature as visual images—that is, composed of large numbers of separate spots. However, I suspect that this is an illusion, because those mental images do not behave in the ways that pictures do. Instead, it seems more likely that we can envision such scenes by intervening at higher levels of the representations that we described in the previous section.

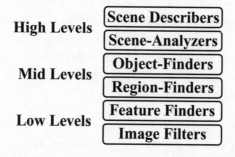

**Making changes at very low levels:** In principle, we could make a new image by changing each spot of the original picture. This would involve a huge computation, and, if you wanted to shift your point of view, you'd have to compute the whole image again. Besides, to make such changes, you would first need some higher-level representations of what those new images should depict. But then, if those higher-level descriptions can answer your questions, you won't need to compute those images!

**Making changes at intermediate stages:** So rather than changing the picture itself, one could change parts of higher-level descriptions. For example, at the Region-Finder level, one could change the name of that top block's front face from *rectangle* to *triangle*. However, this would cause trouble at other levels, because that triangle's edges would not have the proper relations to the edges of the neighboring shapes.

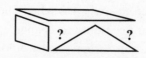

Citizen: When I try to imagine a triangle in my mind, I "see" its three lines as nebulous streaks whose ends don't meet. When I try to correct this by "pushing" a line, it starts to move with some constant speed that I cannot change, and I can't make it stop—yet, strangely, it never gets far away.

That person is trying to change a description but can't maintain some relationships between its parts. When you change an internal representation, it may not maintain its consistency. A real object can't move with two speeds at once, nor can two real lines both meet and not meet. However, imagined objects have fewer constraints.

**Making changes at the higher semantic levels:** You can sometimes avoid such problems by replacing an entire object at some higher level. For example, you might imagine replacing the top of that arch, simply by changing the name of its shape from *rectangular* to *triangular*—if you represent those structures with networks like these, which describe the relationships between their parts.

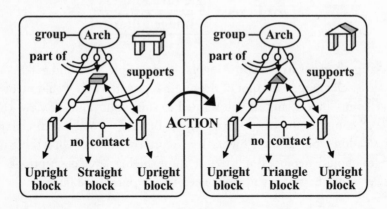

THINKING OF CHANGING THE TOP OF AN ARCH

Chapter 8-7 will say more about such representations, which are sometimes called *"Semantic Networks."* Consider how efficient it is to describe things in terms of such networks! To make such a change at the pictorial level, you would need to change a great number of "pixels"—the separate spots that make up a picture—whereas when you work at a linguistic or other symbolic level, you need only to change a single word or symbol. Making such a change at an earlier stage would involve so many small details that it would be hard to change any part. But at higher "semantic" levels, it is easy to make a meaningful change because, for example, when you describe *"a lying-down block supported by two upright blocks,"* you don't need to mention the viewer's perspective, or even to say which parts of the scene are in view. Thus, the very same description would apply to all these different views:

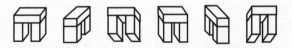

SIX DIFFERENT VIEWS OF AN OBJECT

If we substitute the word *object* for *block,* the same network would also describe other structures like these.

This illustrates the power and efficiency of using more abstract, high-level descriptions; in this case, a word is worth a thousand pictures! In everyday language, the word *abstract* is sometimes used to mean "very hard to understand"—but here it has almost the opposite sense: abstract descriptions are *simpler* because they suppress details that are not relevant.

All this suggests that, in every realm, we can choose to imagine at various levels. Perhaps some chefs imagine new textures and tastes by changing their lower-level sensory states—and some composers do the same with their pitches and timbres—but the same artists might also achieve even better effects with smaller adjustments, selectively made at higher levels.

It seems to me that this subject is important enough that psychologists need a term to describe the various levels at which people construct synthetic perceptions inside their heads—and for this, I have coined the word *simulus* by combining *stimulus* and *simulate.* Thus in Chapter 3-8, I described how I used a simulus of Professor Challenger to disturb me enough to keep myself from falling asleep. To accomplish this, one might try to imagine all the details of such a scene; however, it might be enough merely to represent the high-level abstraction that there was a sneer on my rival's face—without constructing any other low-level details of that imagined simulus.

Drama critic: I can recollect just how I felt while I was attending a certain performance, but I can't remember any details at all of what that dreadful play was about.

Visualizer: When I think about my cat, its image is filled with so many details that I can visualize every hair. Surely there must be advantages to making more realistic picturelike images?[9]

Perhaps when you first imagine that cat, its surface has only a "furry texture"—and only when you "zoom in" on it do you add more details to your mental representation. However, this could happen so rapidly that you have no sense of its happening, and then it may seem to you as though you saw all those details at once. This could be an example of the illusion we mentioned in Chapter 4:

*The Immanence Illusion:* When your questions get answered before you have asked them, it will seem that you're already aware of those answers.

The Immanence Illusion applies not only to *imagined* scenes; we never see *real* scenes "all at once," either. In fact, we don't perceive most fine details until some parts of our minds make requests for them; only then do our eyes turn to focus on them. Indeed, recent experiments demonstrate that our inner descriptions of visual scenes do not get updated continuously.[10]

Consider that in the physical realm, when you think of grasping and lifting a block, you anticipate the feel of its weight—and predict that if you weaken your grasp, then the block will be likely to fall. In the economic realm, if you pay for a purchase, then you will own the thing you have bought, but you must otherwise give it back. In the realm of communication, when you make a statement, then your listeners may remember it—but this will be more likely to happen if you also tell them that it is important.

Every adult knows many such things and regards them to be obvious, but every child takes years to learn how things behave in different realms. For example, if you move an object in the *physical* realm, then this will change that object's location—but if you give some information to your friend, then that knowledge will now be in two places at once. Chapters 6 and 8 will look more closely at how we use such kinds of commonsense knowledge, and will describe a scheme called "Panalogy," which might help to explain how our brains get such answers so rapidly.

## 5-9 Prediction Machines

William James 1890: "Try to feel as if you were crooking your finger, whilst keeping it straight. In a minute it will fairly tingle with the imaginary change of position; yet it will not sensibly move, because *'it is not really moving'* is also a part of what you have in mind. Drop this idea, think of the movement purely and simply, with all brakes off; and, presto! It takes place with no effort at all."

Everyone can think about things without performing any external actions—as when Carol imagined moving those blocks before she actually built anything. But how did she manage to do that? You, yourself, could now close your eyes, lean back in your chair, and indulge in some fantasies and dreams, reflect upon your motives and goals, and then try to decide what next to do.

But how could any brain or machine envision a sequence of possible actions? Section 5-1 showed how to make predictions by using *If* + *Action* → *Then* rules, so a brain could use each *Then* to convert that prediction into a simulus—a representation of the resulting scene—by making a change at some level of our machine's perceptual system. This diagram shows some machinery that could do that kind of processing.

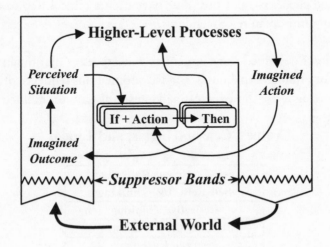

A PREDICTING MACHINE

There are two reasons to include that pair of *"Suppressor Bands."* First, while you are imagining a future condition, you do not want this to be replaced by a description of the present condition; also, you don't yet want your muscles to perform the imagined action until you have considered some other options. So you need some way to disconnect your mind, to enable you to "stop and think" before selecting which action to take.[11] (This could use the same machinery that disconnects our minds from our bodies when we are dreaming while we sleep.)

By repeating its cycle of operation, such a machine could look further into the future, by using the searching and planning schemes that we described in Section 5-3. Furthermore, if it were given enough additional resources, such a mind-machine could simulate what might happen in a larger-scale "virtual world" or indulge in what we call fantasies. Of course, this will need more memory, as well as other kinds of machinery. However, anyone who has played a modern computer game can see that programmers already are quite highly advanced in the art of simulating whole worlds in machines.

Since people clearly do such things, I expect that over the next few years, our scientists will uncover such "prediction machines" in various parts of human brains. How did we evolve these abilities? The species of primates that preceded us must already have had some structures like these, which enabled them to think a few steps ahead. Then, just a few million years ago, those parts of our brains must have rapidly grown, both in size and capacity—and this must have been a critical step toward the growth of our human resourcefulness.

This chapter described some structures and processes that might explain some of our human capabilities, and we outlined this sequence of levels at which we might use those increasingly powerful ways to think.

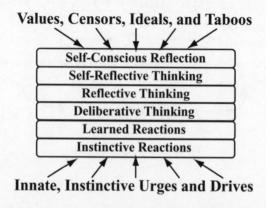

**Values, Censors, Ideals, and Taboos**

| Self-Conscious Reflection |
| Self-Reflective Thinking |
| Reflective Thinking |
| Deliberative Thinking |
| Learned Reactions |
| Instinctive Reactions |

**Innate, Instinctive Urges and Drives**

OUR SIX-LEVEL MODEL OF MIND

However, no matter how such a system is built, it will never be very resourceful until it knows a great deal about the world it is in. In particular, it must be able to foresee some of the outcomes of possible actions, and it won't be able to do this until it possesses a great deal of what we call "commonsense knowledge" and "reasoning." So the subject of common-sense thinking will be the main concern of Chapter 6.

# 6
# COMMON SENSE

"The way to make money is to buy stock at a low price, then when the price goes up, sell it. If the price doesn't go up, don't buy it."
—*Will Rogers*

Soon after the first computers appeared, their blunders became the subjects of jokes. The tiniest errors in programming could wipe out clients' bank accounts, or send out bills for outlandish amounts, or trap the computers in cyclical loops that kept repeating the same mistakes.[1] This maddening lack of common sense led most of their users to conclude that machines could never become intelligent.

Today, of course, computers do better. Some programs can beat people at chess. Others can diagnose heart attacks. Yet others can recognize pictures of faces, assemble cars in factories, or even pilot ships and planes. But no machine yet can make a bed, or read a book, or babysit.

What makes our computers unable to do the sorts of things that most people can do? Do they need more memory, speed, or complexity? Do they use the wrong kinds of instruction-sets? Do their limitations come from the fact that they use only zeros and ones? Or do machines lack some magical attribute that only a human brain can possess? This chapter will argue that none of those are responsible for the deficiencies of today's machines; instead, all those limitations come from the out-of-date ways in which programmers have chosen to program them.

• **No present-day programs have Commonsense Knowledge.** Each present-day program is equipped with no more knowledge than it needs for solving some particular problem. The first parts of this chapter will discuss the huge amounts of knowledge that people have and the skills that they use to apply it. For example, if someone said that a package was tied up with "string," you would understand "obvious" facts like these— whereas no computer yet knows such things:

> With a string you can pull, but not push, a thing.
> If you pull too hard, a string will break.
> You must fill a package before tying it up.
> Loose strings tend to get tangled and knotted.

• **Present-day programs do not have explicit Goals.** Today, we only tell programs some things to do—without telling them why we want them done. So those programs have no ways to tell if their user's goals were achieved at all—or with what quality and at what cost. The middle parts of this chapter will talk about what goals are and how machines could incorporate them.

> People like to go indoors when it rains. (People do not like to get wet.)
> People don't like to be interrupted. (People want you to listen to them.)
> It is hard to hear in a noisy place. (People want to hear what others say.)
> No one else can tell what you're thinking about. (People value privacy.)

• **Few present-day programs are Resourceful enough.** A typical program will simply quit when it lacks some knowledge it needs, or when the method it's using fails—whereas a person will find other ways to proceed. The final parts of this chapter discuss some tactics we use when we don't know exactly what to do—for example, by making analogies.

> Have I been in such situations before?
> What kinds of problems is this one similar to?
> How did I manage to solve those problems?
> Can I adapt those solutions to work on this problem?

The lack of such abilities is why, when something goes wrong, our computers completely break down and stop—instead of finding something better to do. Why can't they learn from experience? All this is because they lack "common sense"!

We do not often recognize how intricate are the processes that we use in every minute of everyday life. This chapter will show that many "commonsensical" things we do are actually far more complex than are many of the specialized skills that attract more attention and respect.

## 6-1　What Do We Mean by Common Sense?

"Common sense is the collection of prejudices acquired by age eighteen."
　　　　—*Albert Einstein*

Instead of blaming machines for their deficiencies, we should try to endow them with more knowledge about the world they are in. This should include not only what we call "commonsense knowledge"—the kinds of facts and concepts that most of us know—but also the "commonsense reasoning" skills which people use for applying their knowledge.

> Student: Can you more precisely define what you mean by "commonsense knowledge"?

We each use terms like *commonsense* for the things that we expect other people to know and regard as obvious.

> Sociologist: That word has different meanings for each of us because what we regard as obvious depends upon the community in which we were raised—such as one's family, neighborhood, language, clan, nation, religion, school, and profession—each of which shares different collections of knowledge, beliefs, and ways to think.

> Child psychologist: Still, even if you know only a child's age, you can say much about what that child is likely to know. Researchers like Jean Piaget have studied children all over the world and found many realms of thought in which they share very similar sets of ideas and beliefs.

Citizen: We also say that people "lack common sense" when they reason in ways that seem foolish to us—not because they are lacking in knowledge, but because they're not using it properly.

Every person is constantly learning, not only new facts but also new kinds of Ways to Think. We learn some from our private experience, some from the teaching of parents and friends, and some from other people we meet. All this can make it hard to distinguish between what each person happens to know and what others regard as obvious—and this can make it hard to predict how anyone else is likely to think.

## The Telephone Call

"You cannot think about thinking without thinking about thinking about something."
—*Seymour Papert*

We'll start by following Papert's advice—by thinking about some ways to think about this typical commonplace incident:

*Joan heard a ring, so she picked up her phone. Charles was answering a question she asked about a particular chemical process. He advised her to read a certain book, which he will shortly bring to her, since he will be in her neighborhood. Joan thanked him and ended the call. Soon Charles arrived and gave her the book.*

Each phrase of that story evokes in your mind some of these kinds of understandings:

**Joan heard a ring.** She recognizes that this special sound means that someone wishes to speak with her.
**She picked up the phone.** Compelled to respond, she crosses the room and moves the receiver to her ear.
**Charles was answering a question she asked.** Charles is in a different room. They both know how to use telephones.
**He advised her to read a certain book.** Joan understands what Charles has said.

***Joan thanked him.*** Was that just a formality or was she genuinely
grateful to him?

***He'll soon be in her neighborhood.*** Joan won't be surprised when he
arrives.

***He gave her the book.*** We don't know if this was a loan or a gift.

We draw such conclusions so fluently that we don't even know that we are
doing it. So let's try to examine how much is involved when one understands
what happened when Joan heard that sound and picked up that phone.

First, when Joan looks at her telephone, she sees only a single side
of it, yet she feels that she's seeing the entire thing, And even before she
reaches for it, she anticipates how it will fit in her grasp, and how it will
feel when it contacts her ear, and knows that one speaks into *here* and
hears answers from *there*. She knows that if she dials a number, some other
phone will ring somewhere else—and if anyone happens to answer it, then
she and that other person can start to converse.

All this rapid retrieval of knowledge seems a natural aspect of seeing
an object—and yet, one has only detected some patches of light! How
does such scanty evidence make it seem as though what you are "looking
at" has been transported right into your mind—where you can move it
and touch it and turn it around, or even open it up and look inside? The
answer, of course, is that what you "see" does not come from your vision
alone, but also comes from how those visual clues lead you to retrieve
other knowledge.

However, on the other side, you know so much about such things
that, surely, your mind would be overwhelmed if you had to "attend" to all
that knowledge at once. So our next few sections will discuss how brains
might interconnect fragments of knowledge so that we can often retrieve
just the items that we need.

## The Concept of a "Panalogy"

Douglas Lenat 1998: "If you pluck an isolated sentence from a
book, it will likely lose some or all of its meaning—i.e., if you
show it out of context to someone else, they will likely miss some
or all of its intended significance. Thus, much of the meaning of
a represented piece of information derives from the context in
which the information is encoded and decoded. This can be a

tremendous advantage. To the extent that the two thinking beings
are sharing a common rich context, they may utilize terse signals
to communicate complex thoughts."

Every word, event, idea, or thing can have many different meanings to
us. When you hear *"Charles gave Joan the book,"* that might make you
think of that *book* as a physical object, or as a possession or possible gift.
And you could interpret this "giving act" in at least these three different
realms of thought:

> *The Physical Realm:* Here "give" refers to the book's motion through
> space, as it moves from Charles's hand to Joan's.

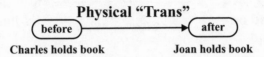

> *The Social Realm:* One might wonder about Charles's motivation. Was
> he just being generous, or was he hoping to ingratiate himself?

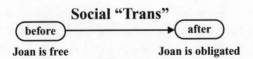

> *The Dominion Realm:* We may infer that Joan is not only holding that
> book, but also has gained permission to use it.

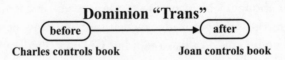

The Dominion Realm is important because you need tools, supplies,
and materials to solve most problems or carry out plans—but most objects
in our civilized world are controlled by persons or organizations that won't
let you use them without their permission.

So here we see three meanings for "give"—which each have somewhat similar structures. Chapter 8-3 will suggest that in such cases, our brains connect analogous items of knowledge from different realms (or different points of view) to the same "roles" or "slots" in one larger-scale structure.

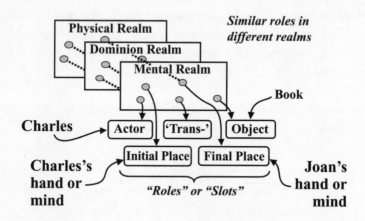

THREE MEANINGS OF "GIVE"

This diagram illustrates a type of structure that I will call a *"panalogy"* (for "parallel analogy"). This scheme makes it easy to rapidly switch between different Ways to Think about an object, idea, or situation. For example, when Joan reaches to grasp and hold that book, she anticipates (in the Physical Realm) the heft of its weight and predicts that if she weakens her grasp, then the book will be likely to fall. Also (in the Dominion Realm) she needs to represent whether she owns it or has to give it back. Thus, if Charles tells her that the book is a gift, then she can cancel that obligation.

What happens when you interpret a certain event in an inappropriate realm of thought? Then often, as soon as you recognize this, you switch to a more useful point of view, without any sense of starting over again. How could you do this so rapidly? Chapter 8 will argue that this could come from your use of panalogies: if you have already linked the same symbol to an appropriate set of multiple meanings, then switching may take no time at all—if the parts of your brain that work in those other realms have already done some processing. This could happen, for example, when you switch from thinking of that book as an object to regarding it as a possession or as a potential collection of knowledge.

More generally, most every concept we entertain has connections to several different realms. For example, the girl in this scene about playing with blocks is likely to have all these kinds of concerns:

*Physical:* What if I pulled out that bottom block?
*Social:* Should I help him with his tower or knock it down?[2]
*Emotional:* How would he react to that?
*Mnemonic:* Where did I leave the small triangular block?
*Strategic:* Can I reach that arch-shaped block from here?
*Visual:* Is the long flat block hidden behind that stack?
*Tactile:* What would it feel like to grab three blocks at once?
*Architectural:* Are there enough extra blocks to make a table?

We again see how a thing or idea can be viewed as having multiple meanings. We sometimes call these "ambiguities" and regard them as defects in how we express or communicate. However, when these are linked into panalogies, then we can think about them in alternative realms—without the need to start over again.

Student: Your example of giving a book suggests that we use the same techniques to represent transportation in space, for transferring an ownership, and for transmitting knowledge to other brains. But what could have led our minds to treat such different ideas in such similar ways?

It surely is no accident that our language uses the same prefix, *"trans,"* in *transfer, transport, transmit, translate, transpose,* etc.—because that common word part, *"trans,"* induces us to make many widely useful analogies.[3] Each of us knows thousands of words, and each time we learn how others use one of them, we inherit another panalogy.

Student: How many different realms can a person use for any particular concept or object? How many of them can we handle at once? How does one know when it's time to switch? To what extent do different persons partition their worlds into similar realms?

More research on semantics will eventually clarify questions like these, but the following sections will discuss only a few of the realms in which one might think about a telephone.

## Subrealms of the Telephone World

We've mentioned only a few of the things that every telephone user knows. However, to *use* what you know about telephones, you also need to know how to speak, and how to understand some of what you may hear. You also need a good deal of knowledge about what people are and how they think, so that you can turn their interests toward the subjects that you want to discuss. Consider how many different knowledge realms we engage to understand the story about Joan's telephone call.

*The Physical Realm:* Joan is close to her telephone, but Charles must be in some more distant place.
*The Dominion Realm:* Both Joan and Charles have telephones, and Charles has dominion over that book. But we can't be quite certain of which objects they own.
*The Procedural Realm:* How does one make a telephone call? We could represent this in terms of a script in which certain actions are specified, but others require you to improvise.

○ *Find telephone number.*
○ *Locate the telephone.*
○ *Pick up phone. Wait for tone.*
○ *Dial number. Wait for ring.*
○ *Initial salutation, e.g., "Hello."*
○ *Specific discussion.*
○ *Terminating salutation.*
○ *Hang up phone.*

First, you must find the phone and dial a number. Then, once the connection has been established, you're supposed to begin with some pleasantries. Eventually, you should say why you called—and then depart from the typical script. At the end you'll close the conversation by saying "good-bye" and "hanging up." Generally, such behavioral scripts begin and end with conventional steps, with improvisations in between. Also, you'll have to depart from the script if something goes wrong—and know how to deal with a wrong connection, or what to do if no one replies, or if you hear the whine of a modem, or if there is too much noise on the line.

**The Social Realm:** When that telephone rings from across the room, Joan will have to walk over to get it; she knows it will do no good to ask, *"Telephone, would you please come here!"* To make an inanimate object move, you have to push, pull, or carry it. But if you want a person to move, those actions would be considered rude; instead, you're expected to make a request. It takes our children quite a few years to learn enough such social rules.

**The Economic Realm:** Every action incurs some cost—not only in materials, time, and energy, but also by closing off alternatives that might bring different benefits. This raises questions about how much effort and time one should spend at comparing the costs of those options. I suspect that there's no simple answer to that, because it depends so much on the present state of the rest of one's mind.

**The Conversational Language Realm:** Most people are experts at dialog, but consider how complex are the skills involved in a typical verbal exchange. You must constantly keep track of the topic, your goal, and your social role. To maintain the respect of your listeners, you must guess what they already know and remember what has already been said so that you won't be too repetitive. It is annoying to be told things one already knows, like *"People can't see the backs of their heads,"* so your conversation must partly be based on your models of what your listeners know about the subjects that are being discussed.

You can communicate your apprehensions and hopes—or try to disguise your intentions; you know that every nuance of expression can strengthen or weaken social bonds; each phrase can persuade or intimidate, conciliate or irritate, or ingratiate or drive away. You also need to keep searching for clues about how well they have understood what you have said—and why you were trying to tell them those things.

> Humanist: Speaking over a telephone is a poor substitute for a face-to-face exchange. The telephone lacks the "personal touch" through which your gestures can put the others at ease, or express the strength of your feelings.

One always loses some nuances when conversing with a person at some other location. On the other side, we're not always aware of the misconceptions that result from what we call "face-to-face" interactions. What if the stranger whom you have just met should resemble (in manner or facial appearance) some trusted friend or some enemy? If that person reminds you of some old Imprimer, this can arouse a misplaced affection or unjustified sense of intimidation. You may think you can later correct such mistakes—but one can never completely erase the "first impression" that one makes.

**The Sensory and Motor Realms:** We also all share many abilities that we don't usually call "commonsensical"—such as the kinds of physical skills that Joan uses to answer that telephone call. It takes less than a single second for you to reach out your arm and *"pick up the phone"*—yet consider how many subgoals this involves:

> Determine the telephone's location.
> Determine its shape and orientation.
> Plan to move your hand to its place.
> Plan how your hand will grasp its shape.
> Plan to transport it toward your face.

Each step of that script raises questions about how we do those things so quickly. We can program computers to do such things, but we do not know how we do them ourselves. It is often supposed that such actions are done under continuous "feedback control"—by processes that keep working to reduce your distance from your goal. However, that cannot be generally true because human reactions are so slow that it takes about one-fifth of a second to react to events that one did not expect. This means that *you cannot change what you are doing right now;* all you can do is revise the plan that you've made for what you will do after that. Thus, when Joan reaches out to answer that call, she must plan to reduce the speed of her hand before it collides with that telephone. Without good plans for what will happen next, she'd be constantly having accidents.

**Kinesthetic, Tactile, and Haptic Realms:** When you squeeze your phone between shoulder and cheek, you anticipate its texture and weight, adjust your grip so that it won't slip, and expect those pressures to disappear as soon as you release it. You already know that this object will fall if released from your grasp, or will break when subjected to too large a stress. An immense amount of such knowledge is stored in your spinal cord, cerebellum, and brain—but those systems are so inaccessible that we can scarcely begin to think about them.

**Cognitive Realms:** We are almost equally inept at describing the systems we use when we think. For example, we are almost completely unaware of how we retrieve and combine the various fragments of knowledge we need—or of how we deal with the risks of being wrong when these involve uncertainties.

**The Self-Knowledge Realm:** Whatever you may be trying to do, you'll need models of your own abilities. Otherwise, you'll set goals that you'll never achieve, make elaborate plans that you won't carry out, or too frequently switch between interests—because, as we'll see in Chapter 9-2, it is hard to achieve any difficult goals unless one can make oneself persist at them.

It would be easy to add to this list of realms, but hard to construct clear distinctions between them.

## 6-2  Commonsense Knowledge and Reasoning

*Robertson Davies 1992:* "You like the mind to be a neat machine equipped to work efficiently, if narrowly, and with no extra bits or useless parts. I like the mind to be a dustbin of scraps of brilliant fabric, odd gems, worthless but fascinating curiosities, tinsel, quaint bits of carving, and a reasonable amount of healthy dirt. Shake the machine and it goes out of order; shake the dustbin and it adjusts itself beautifully to its new position."

I once encountered a fellow professor who was returning from teaching a class, and I asked him how the lecture went. The reply was that it had not gone well because *"I couldn't remember which concepts were hard."* This suggests that, over time, such experts convert some of their high-level skills into lower-level scriptlike processes that leave so few traces in memory that those experts cannot explain how they actually do those things. This has led many thinkers to classify knowledge into two kinds:

*Knowing What.*  This is the kind of "declarative" or "explicit" knowledge that we can express in gestures or words.

*Knowing How.*  These are the kinds of "procedural" or "tacit" skills (like walking or imagining) that we find very hard to describe.

However, this popular distinction doesn't describe the functions of those types of knowledge. It might be better to classify knowledge in terms of the kinds of thinking that we can apply to it:

*Positive Expertise.*  Knowing the situations in which to apply a particular fragment of knowledge.

*Negative Expertise.*  Knowing which actions not to take, because they might make a situation worse.

*Debugging Skills.*  Knowing alternative ways to proceed when our usual methods fail.

*Adaptive Skills.*  Knowing how to adapt old knowledge to new situations.

The first large-scale attempt to catalog commonsense knowledge was the "CYC" project of Douglas Lenat, which started in 1984. Many ideas in this section were inspired by the results of that project.

Douglas Lenat 1998: "In modern America, this encompasses recent history and current affairs, everyday physics, 'household' chemistry, famous books and movies and songs and ads, famous people, nutrition, addition, weather, etc. . . . [It also includes] many 'rules of thumb' largely derived from shared experiences— such as dating, driving, dining, daydreaming, etc.—and human cognitive economics (misremembering, misunderstanding, etc.), and shared modes of reasoning both high (induction, intuition, inspiration, incubation) and low (deductive reasoning, dialectic argument, superficial analogy, pigeon-holing, etc.)."

Then Lenat examines a single sentence: *"Fred told the waiter he wanted some chips,"* to see the kinds of knowledge one might require to understand what that statement means.[4]

The word *he* means Fred—and not the waiter. This event took place in a restaurant. Fred was a customer dining there. Fred and the waiter were a few feet apart. The waiter was at work there, waiting on Fred at that time.

Fred wants potato chips, not wood chips. Fred does not want some particular set of chips.

Fred accomplished this by speaking words to the waiter. Both Fred and the waiter are live human beings. Both of them speak the same language. Both were old enough to talk, and the waiter was old enough to work.

Fred is hungry. He wants and expects that in a few minutes the waiter will bring him a typical portion—which Fred will start eating soon after he gets them.

We can also assume that Fred assumes that the waiter also assumes all those things.

Here is another example of what one must know to understand a commonplace statement:

"Joe's daughter was sick so he called the doctor."

We can assume that Joe cares about his daughter, is upset because she is sick, and wants her to be healthy. Presumably he believes she is sick because of observing some symptoms.

People have different abilities. Joe himself cannot help his daughter. People ask others for help to do things they can't do themselves. So Joe called the doctor to help heal his daughter.

Joe's daughter, in some sense, belongs to Joe. People care more about their own daughters than about other people's daughters. If so advised, Joe will take the daughter to the doctor. When at the doctor's, she will still belong to Joe.

Medical services can be expensive, but Joe is likely to forgo other spending to get the doctor to help the daughter.

All these are things that "everyone knows" and uses to understand everyday stories. But none of that knowledge would have much use unless we also had additional knowledge about which fragments of knowledge might help us to achieve each particular kind of goal.

## How Much Does a Typical Person Know?

"A little knowledge is a dangerous thing. So is a lot."
—*Albert Einstein*

Everyone knows a good deal about many objects, topics, and ideas—and this might lead one to suppose that we each have enormous memories. And many writers have argued that, since each human brain has trillions of synapses, then surely we must use these to store at least many billions of memories. However, if the arguments in this section are right, then our bodies of knowledge might not be so vast.

In any case, let's start by making a minimal estimate. We know each person knows thousands of words, and it seems safe to assume that a typical word might be linked in our minds to perhaps a thousand other items of memory. This means that a person's language system might have the order of a few million links. Similarly, in the physical realm, we each know thousands of kinds of objects—and typically, each might be linked to a thousand other objects and uses. Similarly, in the social realm, you might know thousands of things about each of a hundred people and hundreds of things about each of a thousand people.

This suggests that in each important realm, one might know perhaps a few million things. But while it is easy to think of a dozen such realms, it is hard to think of a hundred of them. So all this suggests that, for a machine to do humanlike reasoning, this might require only the order of a few hundred millions of items of knowledge.[5]

> Citizen: Perhaps so, but I have heard of phenomenal feats of memory. What about persons with photographic memories, who can recollect all the words of a book after only a single reading of it? Could it be that we all remember, to some extent, everything that happens to us?

We all have heard such anecdotes, but whenever we try to investigate one, we usually fail to uncover the source, or find that someone was fooled by a magic-show trick. We sometimes encounter a person who has memorized the contents of a few sizable books—but I have heard of no rigorous demonstration that someone has memorized a hundred such books.[6] Here is what one psychologist said about a person who appeared to possess a prodigious memory:

> Alexander R. Luria 1968: "For almost thirty years [I] had an opportunity systematically to observe a man whose remarkable memory . . . for all practical purposes was inexhaustible. . . . It was of no consequence to him whether the series I gave him contained meaningful words or nonsense syllables, numbers or sounds; whether they were presented orally or in writing. All that he required was that there be a three-to-four-second pause between each element in the series. . . . And he could manage, also, to repeat the performance fifteen years later, from memory."

This performance may seem remarkable, but might not be truly exceptional, because Thomas Landauer (1986) concluded that, during any extended interval, none of his subjects could learn at a rate of more than about two bits per second, whether the realm be visual, verbal, musical, or whatever. So if Luria's subject required a few seconds per word, his performance would fit Landauer's estimate.[7]

> Student: I'm uncomfortable with this argument. I agree that it might apply to our higher-level kinds of knowledge. But our sensory and motor skills might be based on much larger amounts of information.

We don't have a good way to measure such things, and making such estimates raises hard questions about how those fragments of knowledge are stored and connected. Still, we have no solid evidence that any person has ever surpassed the limits that Landauer's research suggests.[8]

Chapter 7 will speculate about how we organize knowledge so that, whenever one of our processes fails, we can usually find an alternative. But here we'll change the subject to ask how we could endow a machine with the kinds of knowledge that people have.

## Could We Build a "Baby-Machine"?

Here is an old and popular dream: to build a machine that starts by learning in simple ways and then later develops more powerful methods—until it becomes intelligent.

> Entrepreneur: Why not build a "baby-machine" that learns what it needs from experience? Equip a robot with sensors and motors, and program it so that it can learn by interacting with the real world—the way that a human infant does. It could start with simple *If→Then* schemes, and then later invent more elaborate ones.

In fact, several actual projects have had this goal, and each such system made progress at first but eventually stopped extending itself.[9] I suspect that this usually happened because those programs failed to develop *good new ways to represent knowledge*. Indeed, inventing good ways to represent knowledge has long been a major goal in computer science. However, even when new ones are discovered, they rarely are quickly and widely adopted—because one must also develop good skills to work with them efficiently. And since such skills take time to grow, their users will need to tolerate extensive periods during which their performance becomes not better, but worse.[10] (See Section 6-7 and Chapter 9-4.) In any case, no one has yet made a baby-machine that was able to keep on developing effective new kinds of representations.

Another problem with baby-machines is that if a system learns new rules too recklessly, it is likely to accumulate too much irrelevant information—and its performance will deteriorate. Chapter 8-5 will argue that unless learning is done selectively—by making appropriate "credit assignments"—a machine will fail to learn the right things from most of its experiences.

> Entrepreneur: Instead of trying to build a system that learns by itself, why not make one that searches the Web to extract knowledge from those millions of pages of content-rich text.

That certainly is a tempting idea, for the World Wide Web must contain more knowledge than any one person ever could learn. However, the texts on the Web do not *explicitly* display the knowledge that one would need to understand what all those texts mean.[11] Thus, consider the kind of story we find in a typical child's reading book:

> "Mary was invited to Jack's party. She wondered if he would like a kite. She went and shook her piggy bank. It made no sound."

A typical reader would assume that Jack is having a *birthday party,* and that Mary is concerned about the need to bring a *birthday present* to Jack.[12] A good birthday present should be something that its recipient likes—and the suggestion that Jack might like a kite also suggests that Jack is a child, and that a kite might be a suitable toy. Mentioning a piggy bank suggests that Mary is thinking of buying a kite and needs money to pay for it. Also, the bank would have rattled if it contained coins; this implies that Mary now faces a financial problem. However, unless the reader knows all these facts, this "simple" story would make no sense, because there would be no apparent connection between each of its sentences and the next.

> Neurologist: Why not try to copy the brain, using what brain scientists have learned about the functions of various parts of the brain.

We learn more about such details every week—but still do not yet know enough to simulate even a spider or snake.

Programmer: What about alternatives such as building very large
machines that accumulate huge libraries of statistical data?

Such systems can learn to do useful things, but I would expect them to
never develop much cleverness, because they use numerical ways to represent all the knowledge they get. So until we equip them with higher
reflective levels, they won't be able to represent the concepts they'd need
for understanding what those numbers might mean.

Evolutionist: If we don't know how to design better baby-machines,
perhaps we can make them evolve by themselves. We could first
write a program that writes other programs and then makes various
kinds of mutations of them—and then make all those programs
compete for survival in suitably lifelike environments.

It took hundreds of millions of years for us to evolve from the earliest ver-
tebrate fish, and it took all of those eons for us to develop the structures
that became the higher reflective levels that we described in Chapter 5.
The following chapters will argue that each human child makes extensive
use of those high-level structures to develop our uniquely human ways to
represent new kinds of knowledge and processes. It seems clear to me that
this is why the attempts to make baby-machines have led to unimpressive
results: *you cannot learn things that you can't represent.*

John McCarthy 1959: "If one wants a machine to be able to discover an abstraction, it seems most likely that the machine must be
able to represent this abstraction in some relatively simple way."

I do not mean to dismiss all prospects of building a baby-machine, but I
suspect that any such system would develop too slowly unless (or until) it
was equipped with adequate ways to represent knowledge (see Chapter 8).
In any case, it seems fairly clear that human brains are innately equipped
with highly developed ways to learn (some of which don't start to operate until long after birth). The researchers who have tried to build such
machines have used quite a few ingenious schemes, but it seems to me that
each of those machines got stuck because of not having ways to overcome
one or more problems like these:

*The Optimization Paradox:* The better a system already works, the more likely each change will make it worse—so it gets more difficult for it to find more ways to improve itself.

*The Investment Principle:* The better a certain process works, the more we will tend to rely on it, and the less we will be further inclined to develop new alternatives—especially when a new technique won't yield good results until you become proficient with it.

*The Complexity Barrier:* The more that the parts of a system interact, the more likely each change will have unexpected side effects.

Evolution is often described as a process of *selecting* beneficial changes—but most of evolution's work involves *rejecting* changes that have bad effects! This surely is why most species evolve to occupy narrow, specialized niches that are bounded by all sorts of hazards and traps. It is not often recognized that while genetic evolution can "learn" to avoid the most common kinds of mistakes, it is virtually incapable of learning large numbers of very uncommon mistakes. Indeed, only a few "higher animals" have escaped from this by evolving language-like systems through which they can inform their descendants about accidents that happened to some of their ancestors' relatives.

All this suggests that it will be difficult for a machine to keep developing—unless it first evolves ways to protect itself against changes that cause bad side effects. An excellent way to accomplish this, both in engineering and in biology, has been to split the whole system into parts that then can evolve more independently. This surely is why all living things evolved to become assemblies of separate parts (which we call *"organs"*)—each of which have comparatively few connections to other parts.

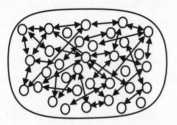
AN UNSTRUCTURED SYSTEM

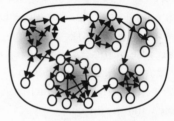
AN "ORGANIZED" SYSTEM

In an organ-based structure, a change in one organ will have fewer bad effects on what happens inside the other organs. In particular, this could be why the resources inside our brains evolved to become *"organ-ized"* into more-or-less separate centers and levels.

> Alan Turing 1950: "We cannot expect to find a good child machine at the first attempt. One must experiment with teaching one such machine and see how well it learns. One can then try another and see if it is better or worse [but] survival of the fittest is a slow method for measuring advantages. The experimenter, by the exercise of intelligence, should be able to speed it up [because] if he can trace a cause for some weakness he can probably think of the kind of mutation which will improve it."

## Remembering

Whenever we get a new idea, or find a new way to solve a problem, then we may make a memory record of it. But records are useless unless you have ways to *"re-collect"* the ones that are relevant to the problem that you are facing right now. I'll argue that this needs a lot of machinery.

> Citizen: If remembering is so complex, then why does it seem so effortless, simple, and natural? Each idea reminds me of similar ones, which then make me think of related ideas—until I recall the ones that I need.

Why does "remembering" seem so effortless? As long ago as you can remember, you could always recall things that happened to you. However, you cannot remember much of your earliest years; in particular, you cannot recall how you developed your early abilities. Presumably, you had not yet developed the skills one needs for making those kinds of memories. (See Johnston 1997.)

Because of this *Amnesia of Infancy,* we all grow up with simplistic views of what memories are and how they work. You might think of your memory as like a writing pad, on which you can jot down mental notes. Or perhaps for each significant event, you store "it" away in some kind of memory-box and later, when you want it back, you somehow bring "it" out of that box—if you are lucky enough to find it. But what kinds of structures do we use to represent those "its" and how do we bring them back when we need them? Our recollections would be useless unless (1) they were relevant to our present goals and (2) we also had ways to retrieve the ones that we need at the times when we need them.

To retrieve information rapidly, a computer expert might suggest that we store everything in some single "database" and use some general-purpose "matching" technique. However, most such systems still classify things in terms of *how those things have been described* instead of in terms of *the goals that they can help us to achieve.* This is extremely important because we usually know less about the *type of thing we are looking for* than about *the goal that we want to accomplish with it*—because we're always facing some obstacles, and want to know how to deal with them.

So instead of using some "general" method, I suspect that every child develops ways to link each new fragment of knowledge to some particular goals it might help to achieve, and thus help to answer questions like these:

> *What kinds of goals might this item serve?*  Which kinds of problems could it help to solve? What obstacles could it help to overcome?
> *In which situations might it be relevant?*  In which contexts is this likely to help? What subgoals must first be achieved?
> *How has it been applied in the past?*  What were some similar previous cases? What other records might be relevant? See *Credit Assignment* in Chapter 8-5.

Each fragment of knowledge may also need links to some knowledge about its deficiencies—and the dangers and costs of using it:

> *What are its most likely side effects?*  Is it likely to do us more harm or more good?
> *How much will it cost to use it?*  Will it repay the effort of using it?
> *What are its common exceptions and bugs?*  In which contexts is it likely to fail us—and what might be good alternatives?

We also link each item to information about its sources and to what certain other persons might know.

*Was it learned from a reliable source?* Some informants may simply
be wrong, while others may mean to mislead us.

*Is it likely to be outdated soon?* That's why this book does not
discuss most current beliefs about how our brains work.

*Which other people are likely to know it?* Our social activities
strongly depend on knowing what others may understand.

All this raises questions about how we make so many connections to and
from each new fragment of knowledge. I suspect that we can't do this all
at once—and indeed there is some evidence that it normally takes some
hours or days (including some sessions of dream-laden sleep) to establish
new long-term memories. Also, we probably add more links each time
we retrieve a fragment of knowledge, because then we're likely to ask ourselves, *"How did this item help (or hinder) me at overcoming this obstacle?"*
Indeed, some research in recent years suggests that our so-called long-term
memories are not so permanent as we used to think; it seems that they can
be altered by suggestions and other experiences.

We all know that our memory systems can fail. There are things that we
can't remember at all. And sometimes we tend to recollect not what actually
happened to us but versions that seem more plausible. At other times we fail
to remember something relevant until—after several minutes or days—suddenly the answer appears, and you say to yourself, *"How stupid of me; I knew
that all along!"* (That could happen either because an existing record took
long to retrieve, or because it was never actually there—and you had to construct a new idea by using some process of reasoning.)

In any case, we should expect such "lapses" because our recollections
*must* be selective; Chapter 4-4 discussed how bad it would be to remember
everything all the time: it would overwhelm us to recall all the millions of
things that we know. However, none of this answers the question of how
we usually retrieve the knowledge that we currently need. I suspect that we
do this mainly by our already having prepared in advance the sort of links
discussed above. But constructing these requires additional skills, which
we'll discuss in Chapter 8-5.

At the start of this section we asked about how we retrieve the knowledge we need. The following section will argue that part of the answer
lies in those links to *the goals that each fragment of knowledge might help to
achieve.* To make that statement more concrete, the next few sections will
investigate what goals are and how they work.

## 6-3  Intentions and Goals

> Alan Watts 1960: "No one imagines that a symphony is supposed
> to improve in quality as it goes along, or that the whole object of
> playing it is to reach the finale. The point of music is discovered in
> every moment of playing and listening to it. It is the same, I feel,
> with the greater part of our lives, and if we are unduly absorbed in
> improving them we may forget altogether to live them."

Sometimes we seem to act passively, just reacting to things that happen to
us—but at other times we feel more in control, and feel that we're actively
choosing our goals. I suspect that this most often happens when two or
more goals become active at once and thereby lead to a conflict. For as we
noted in Chapter 4-1, when our routine thinking runs into trouble, this
engages our higher reflective levels.

For example, when angry or greedy enough, we are likely to take
actions that later may make us have feelings of shame or guilt. Then we
may offer such justifications as, *"That impulse became too strong to resist"* or
*"I found that I did it in spite of myself."* Such excuses relate to the conflicts
between our immediate goals and our higher ideals, and every society tries
to teach its members to resist their urges to breach its conventions. We call
this developing "self-control" (see Chapter 9-2), and each culture makes
maxims about such feelings.

> Moralist: No merit comes from actions based on self-serving wishes.
> Psychiatrist: One must learn to control one's unconscious desires.
> Jurist: To be guilty in the first degree, an offense must be deliberate.

Still, an offender can object, *"I didn't intend to do those things,"* as though a
person is not "responsible" for an action that wasn't intentional. But what
kinds of behavior might lead you to think that a person did something
"deliberately"—in contrast to it having resulted from mental processes
that were not under that person's control?

To understand this, it may help to observe that we have similar thoughts
about physical things; when we find that some object is hard to control, we
sometimes imagine that *it* has a goal—and say, *"This puzzle piece doesn't want
to fit in,"* or *"My car seems determined not to start."* Why would we think of
an object in that way, when we know that it has no such intentions?

The same thing can happen inside your mind, when one of your goals becomes so strong that it is hard to think about anything else. Then it may seem to come from no choice of your own, yet is somehow being imposed upon you. But what could make you pursue a goal that does not seem to be one that you want? This could happen when that particular goal conflicts with some of your high-level values, or when you have other goals with different aims; in any case, there is no reason to expect all of your goals to be consistent.

However, this still does not answer the question of why a goal can seem like a physical force, as in, *"That urge became irresistible."* And indeed, a "powerful" goal can seem to push other goals aside, and even when you try to oppose it, it may prevail if you don't fight back strongly enough. Thus both forces and goals share some features like these:

Both seem to aim in a certain direction.
Both "push back" when we try to deflect them.
Each seems to have a "strength" or "intensity."
Both tend to persist till the cause of them ends.

For example, suppose that some external force is applied to your arm—say, strongly enough to cause some pain—and your *A*-Brain reacts by pushing back (or by moving away)—but, whatever you do, it keeps pressing on you. In such a case, your *B*-Brain might see nothing more than a sequence of separate events. However, your higher reflective levels might recognize these as matching this particular pattern:

"Something is resisting my efforts to make it stop. I recognize this as
a process that shows some persistence, aim, and resourcefulness."

Furthermore, you might recognize a similar pattern inside your mind when some resources make choices in ways that the rest of your mind cannot control, as when you do something "in spite of yourself." Again, that pattern may seem as though some external force was imposed on you. So it often makes practical sense to represent intentions as though they were like forces or even like antagonists.

Student: But isn't it merely a metaphor, to speak of a goal as resembling a force? Surely it's bad to use the same words for things with such different characteristics.

We should never say "merely" for metaphors, because that is what all descriptions are; we can never state just what something *is,* we can only describe what something is *like*—that is, to describe it in terms of other things that seem to us to have some similar properties—and then to consider the differences. Then we label it with the same or a similar name so that thenceforth that older word or phrase will include this additional meaning-sense. This is why most of our words are "suitcase-words." Chapter 9-4 will argue that the ambiguities of our words may be the greatest treasures that we inherit from our ancestors.

We've mentioned goals many times in this book—but never discussed how goals might work. So let us turn from the subject of how a goal feels to ask what a goal might actually be!

## Difference-Engines

> Aristotle a: "Differences arise when what we get is different from what we desire; for it is like getting nothing at all when we do not get what we aim at."

Sometimes people appear to behave as though they had no direction or aim. At other times they seem to have goals. But what *is* a goal, and how can we *have* one? If you try to answer such questions in everyday words like *"a goal is a thing that one wants to achieve,"* you will find yourself going in circles because, then, you must ask what *wanting* is—and then you find that you're trying to describe this in terms of other words like *motive, desire, purpose, aim, hope, aspire, yearn,* and *crave.*

More generally, you get caught in this trap whenever you try to describe a state of mind in terms of other psychology-words, because these never lead to talking about the underlying machinery. However, we can break out of that with a statement like this:

> A system will seem to have a goal when it persists at applying different techniques until the present situation changes into a certain other condition.

This takes us out of the psychological realm by leading us to ask about what kind of machinery could do such things. Here is one way such a process might work:

*Aim:* It begins with a description of a certain possible future situation. It also can recognize some differences between the situation it now is in and that "certain other condition."

*Resourcefulness:* It is also equipped with some methods that may be able to reduce those particular kinds of differences.

*Persistence:* If this process keeps applying those methods, then, in psychological terms, we will perceive it as trying to change what it now has into what it "wants."

Persistence, aim, and resourcefulness! The next few sections will argue that this particular triplet of properties could explain the functions of what we call "motives" and "goals," by giving us answers to the questions that we asked in Chapter 2-2:

What makes some goals strong and others weak?
What are the feelings that accompany them?
What could make an impulse "too strong to resist"?
What makes certain goals "active" now?
What determines how long they'll persist?

No machine had clearly displayed those three traits of *aim, persistence,* and *resourcefulness*—until 1957, when Allen Newell, Clifford Shaw, and Herbert Simon developed a computer program called the *"General Problem Solver."* Here is a simplified version of how it worked; we'll call this version a *"Difference-Engine."*[13]

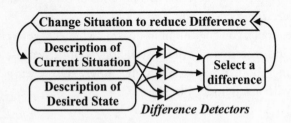

A "DIFFERENCE ENGINE"

At every step, this process compares its descriptions of the present and that future situation, and this produces a list of differences between them. Then it focuses on the most serious difference and applies some technique that has been designed to reduce this particular type of difference. If this succeeds, the program then tries to reduce what *now* seems to be the most serious difference. However, whenever such a step makes things worse, the system goes back and tries a different technique.

Thus, as we mentioned in Chapter 2-2, every infant is born with two systems for maintaining "normal" body temperature: when too hot, the baby may sweat, pant, stretch out, and/or vasodilate.

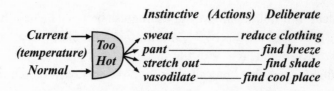

**WAYS TO REACT TO BEING TOO HOT**

However, when the baby is too cold, it will curl up, shiver, vasoconstrict and/or raise its metabolic rate.

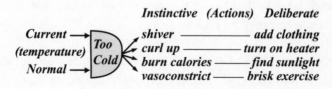

**WAYS TO REACT TO BEING TOO COLD**

At first we may be unaware of such processes, because such instinctive reactions begin at very low cognitive levels. For example, when you become too hot, you automatically start to sweat. However, when perspiration drips, you may notice this, and deliberate: *"I must find some way to escape from this heat."* Then your acquired knowledge may suggest other actions to take, such as moving to an air-conditioned place. If you feel too cold, you might put on a sweater, turn on a stove, or begin to exercise (which can make you produce ten times as much heat). Now we can interpret "having a goal" to mean that a Difference-Engine is actively working to remove those differences.

Student: To have a goal, does one really need a representation of the desired situation? Would it not be sufficient just to have a list of desired properties?

This is a matter of degree, because one could never specify every aspect of a situation. We could represent a "desired situation" as a simple, rough sketch of a future scene, as a list of a few of its properties, or as just some single property (for example, that it is causing some pain).

Student: But still, should we not distinguish between merely "having a goal" and more actively "wanting" it? I would say that your Difference-Engine is a "wanting-machine" and that the goal itself is only the part that you called its "aim"—its current description of some future situation.

I agree that this student is perfectly right: the word *goal* has two different meanings in everyday language. A potential goal becomes an active goal when one is running a process that changes things till they fit that description—and perhaps our everyday language does not help to make the distinctions we need; this is why each specialized field needs to develop its own special "jargon." But I don't think we'll have any trouble, here, about what we mean by "goal" in each context.

Romanticist: This Difference-Engine idea could account for some of what "having a goal" might mean—but it doesn't explain the joy of success, or the distress that we feel when we fail to achieve what we've hoped for.

I agree that no single meaning of *goal* can explain all of those cascades of feelings, because *wanting* is such a large suitcase of concepts that no single idea can embrace them all. Besides, many things that people do come from processes with no goals at all, or goals of which they are unaware. Nevertheless, the Difference-Engine's characteristics capture more of our everyday concept of *goal* than any other description I've seen.

Student: What happens when that Difference-Engine finds several differences at once? Can it work on them all simultaneously, or must it deal with them one by one?

The inventors of the *General Problem Solver* concluded that, when several differences are detected, the machine should first try to remove the most significant difference because this is likely to make a large change in the situation (hence, it might be a waste of time first to remove any smaller differences). To do this, *General Problem Solver* had to assign different priorities to each kind of difference that it could detect.

> Student: What if reducing one of those differences makes several
> other differences worse? This could happen if Carol moves a block
> to a place that prevents her from building the rest of her arch.

When any action makes the largest difference worse, then one may need to search several steps ahead—for example, by using the methods described in Chapter 5-3. However, without such machinery for making plans, a Difference-Engine by itself cannot get past a short-term loss to achieve a larger future gain.

Apparently, this limitation led Newell and Simon to move in other research directions, as seen in Newell 1972. I think they should have persisted, by adding more reflective levels to the basic Difference-Engine scheme—because one could argue that the system got stuck because it was not equipped with ways to reflect on its own performance, the way that people can "stop to think" about the methods that they have been using. Indeed, in a brilliant but rarely cited essay, Newell, Shaw and Simon 1960b themselves suggested an ingenious way to make one Difference-Engine reflect on (and improve) the performance of a second one. However, it seems that no researchers (including them) ever went on to further develop that scheme.

What if one fails to solve a problem, even after using reflection and planning? Then one may start to consider that this goal may not be worth the effort it needs—and this kind of frustration then can lead one to "self-consciously" think about which goals one "really" wants to achieve. Of course, if one elevates that level of thought too much, then one might start to ask questions like, *"Why should I have any goals at all,"* or, *"What purpose does having a purpose serve"*—the troublesome kinds of questions that our so-called "existentialists" could never find plausible answers to.

However, the obvious answer is that this is not a matter of personal choice: we have goals because that's how our brains evolved: the people without goals became extinct because they simply could not compete.[14]

## Goals and Subgoals

Aristotle a: "We deliberate not about ends, but about means. . . .
We assume the end and think *about* by what means we can attain
it. If it can be produced by several means, we consider which one
of them would be best . . . [and then] we consider by which means
*that* one can be achieved, until we come to the first cause (which
we will discover last)."[15]

Chapter 2-2 considered some questions about how we connect our subgoals to goals—but did not stop to investigate how those subgoals might
originate. However, a Difference-Engine does this by itself, *because every
difference it needs to reduce becomes another subgoal for it!* For example,
if Joan is in Boston today, but wants to present a proposal in New York
tomorrow, then she will have to reduce these differences:

The meeting is two hundred miles away.
Her presentation is not yet complete.
She must pay for transportation, etc.

That distance difference is too large to walk, but Joan could drive a car or
travel by train—and she also knows some "scripts" for taking an airplane
trip like this.

> **Get to the airport.**
> **Buy ticket, go to the gate.**
> **Wait on long security line.**
> **Board the plane.**
> **Fly to destination airport.**
> **Local travel to final destination.**

SCRIPT FOR AN AIRPLANE TRIP

However, each phase of this script needs several steps. She could *"Get
to the airport"* by bicycle, taxicab, or bus, but she decides to drive her
car—which itself has a script that has subgoals like these.

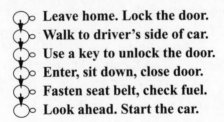

**Leave home. Lock the door.**
**Walk to driver's side of car.**
**Use a key to unlock the door.**
**Enter, sit down, close door.**
**Fasten seat belt, check fuel.**
**Look ahead. Start the car.**

SCRIPT FOR STARTING A TRIP BY CAR

When Joan reviews that airplane trip, she decides it would waste too much of her time to park the car and pass through the security line. The actual flight from her home to New York takes no more than an hour or so, and the railroad trip is four hours long, but it ends near her destination—and she could spend all that time at productive work. She "changes her mind" to take the train.

Similarly, if Carol were to decide to build a tower with her blocks, she would need to divide that job into parts and to make a plan that uses a procedure like this.

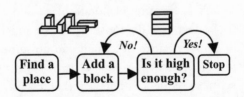

PROCESS FOR BUILDING A TOWER

Then each of those subgoals will turn out to require several more parts and processes—and when we developed a robot to do such things, its software needed several hundred parts. For example, *"Add a block"* needed a branching network of subgoals like this.

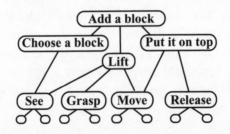

Of course, each subgoal itself may be quite complex. *"Choose a block"* must avoid selecting blocks that already support the tower top. *"See"* must recognize objects regardless of color, size, and shades of light—and even when they are partly obscured by other blocks. *"Grasp"* must adapt the robot's hand to the size and shape of the block to be moved. And *"Move"* must guide the arm and hand through paths that do not strike the tower's top or the child's face.

How does a person discover which subgoals are required to accomplish a job? You could find these by trial and error, or by doing experiments in your head, or by recalling some prior experience—and one of the most useful methods of all is to use a Difference-Engine, because each difference becomes a new subgoal for you.

To summarize, our idea is that to have an *active* goal amounts to running a Difference-Engine–like process. I suspect that, inside each human brain, many such processes all run at once, at various levels in various realms. These range from reactive systems that work all the time (like those that maintain our temperatures) up to the self-conscious levels at which we less frequently think about what kind of person we'd like to be.[16]

How often do people actually use the kinds of techniques that we've been describing—such as forming elaborate plans and splitting up jobs into smaller ones? In fact, we do most things in much simpler ways, because we already know what to do: when you do something a number of times (as when you "practice" a new type of skill), it gradually gets converted into a script or sequence of actions that less often require higher levels of thinking.

> "An expert is one who does not have to think. He knows."
> —*Frank Lloyd Wright*

The result is that we need those search-and-planning techniques only when we face a new kind of problem (or fail to recognize it as a familiar one). But how might "practice" improve a skill, to produce those "expert" performances? An ancient theory of this was the idea that each time you use a "path in the brain," it deepens some sort of memory groove, so that it will be easier to follow that path in the future. A more modern version of this is that the synapses between the cells of the brain become better conductors when they are more used—and there surely must be some truth to this.

However, Chapter 8 of this book will suggest some higher-level ways in which "practice" could improve a performance. For example, some processes might replace an extensive search by a straightforward script that contains only the steps that led to success; in other words, one learns to use a particular path instead of searching through a map. Other processes use repeated attempts to replace the *Ifs* of complex rules by ones that react only to relevant features. And yet other processes may construct new Critics and Censors to prevent various sorts of common mistakes.

In any case, as one increases one's proficiency, one may come to feel a sense of mastery, as though one understands an entire complex realm, and can think of it as a single whole. But this can be an illusion that comes from forgetting the effort of learning one's skills and then turning them into efficient but mindless scripts—in short, by replacing the process of "figuring out" by an unreflective reaction-machine. When this happens, it could be one reason why many achievers become less able to teach others to imitate their techniques.

## 6-4  A World of Differences

> Francis Bacon 1620: "Some minds are stronger and more apt to mark the differences of things, others to mark their resemblances. The steady and acute mind can fix its contemplations and dwell and fasten on the subtlest distinctions: the lofty and discursive mind recognizes and puts together the finest and most general resemblances. Both kinds however easily err in excess, by catching the one at gradations, the other at shadows."

Whenever somebody tells you a story, you react less to what each separate sentence means than to how this differs from what you expected—and this also applies to our other perceptions. For example, if you plunge your hand into a bowl of cold water, you'll feel a severe sensation of chill—but soon this will totally disappear, just as a steady pressure on your skin will quickly seem to diminish in strength. It is the same with new odors or tastes, or with the onsets of continuous sounds: at first those sensations may seem intense, but then they rapidly fade away. We have many different names for this, like *accommodation, adaptation, acclimatization, habituation,* or just becoming *accustomed* to things.

Student: This doesn't apply to vision, though. I can look at an object as long as I like, and its image never fades; in fact, I keep seeing more features of it.

Physiologist: In fact, that image would rapidly fade if you could keep from moving your eyes, which normally make small motions that keep changing your retinal images.[17]

Thus, most of our external sensors react only to rather rapid changes in conditions. (However, we also have additional sensors that do not fade away, but keep responding to certain particular harmful conditions.)

Now let's apply the same idea—of a system that "mainly reacts to change"—to a brain with a tower of cognitive levels. This could help to explain some phenomena. For example, after you start a trip on a train, you're aware of the clacking of wheels on the track, but (if that clacking is regular) then you will soon stop noticing this. Perhaps your A-Brain is still processing it, but your B-Brain has stopped reacting to it. It will be much the same for the visual scenes; when the train enters a forest, you'll start seeing trees—but soon you'll start to ignore them. What could cause such meanings to fade?

It's much the same with repeated words; if someone says "rabbit" one hundred times, while trying to focus on what that word means, then that meaning will shortly disappear—or be replaced by some other one. And similarly the same thing happens when you listen to popular music: first you'll hear dozens of nearly identical measures, but the details of these soon fade away and you no longer pay any attention to them. Why don't we object to that repetitiousness?

This could be partly because we tend to interpret such "narratives" in terms of how situations change on successively larger scales of time. In the case of most music, this structure is clear: we begin by grouping separate notes into "measures" of equal length, and we then group these into larger sections, until the whole composition is seen as a storylike structure.[18] We do this in vision and language, too—although with less repetitiousness—by grouping collections of smaller events into multiple levels of events, incidents, episodes, sections, and plots. However, we see this most clearly in musical forms:

*"Feature-Detectors"* recognize pauses, notes, and various other aspects of sounds, such as harmony, tempo, and timbre, etc.

*"Measure-Takers"* group these into chunks. In music, composers make this easy for us by using measures of equal length; this helps us to sense the differences between successive chunks.

*"Phrase- and Theme-Detectors"* then represent larger events and relationships such as, *"This theme goes down and then goes up, and ends with three short, separate notes."*

Then *"Section-Builders"* group these into larger-scale parts, such as, *"These three similar episodes form a sequence that rises in pitch."*[19]

| Vision | Music | Language |
|--------|-------|----------|
| Stories | Pieces | Plots |
| Localities | Movements | Chapters |
| Scenes | Sections | Paragraphs |
| Objects | Themes | Sentences |
| Regions | Phrases | Phrases |
| Groups | Measures | Words |
| Features | Features | Phonemes |

SIMILAR LEVELS IN DIFFERENT REALMS

Finally, our *"Storytellers"* interpret each piece as similar to events in other realms—such as depicting a journey through space and time, or a skirmish among personalities. One special appeal of music is how effectively it can depict what we might call *"abstract emotional scripts"*—stories that seem to be about entities about whom we know nothing at all except that we can recognize their individual characteristics—e.g., *this one is warm and affectionate,* whereas *that one is cold and insensitive.* Then we empathize with how they feel by interpreting those phrases and themes as representing mental conditions like conflict, adventure, surprise, and dismay—as in, *Those horns are attacking the clarinets, but the strings are now trying to calm them down.*

Now suppose that each higher level in the brain mainly reacts to the changes below it, but over some larger scale of time. If so, then when signals repeat at level *A,* the *B*-Brain will have nothing to say. And if the signals that go up to *B* form a sequence that repeats—so that the *B*-Brain keeps seeing a similar pattern—then the *C*-Brain will sense a "constant condition," and thus have nothing to say to the level above it. More generally, we can expect any repetitive signal to tend to partly "anesthetize" the next level above it. So although your foot may continue to tap to the beat of a rhythmic tune, most details of those smaller events eventually will get ignored.

Why might our brains have evolved to work this way? If some condition has been present for long—and nothing bad has happened to you—then it probably poses no danger to you, so you might as well not pay attention to it and apply your resources more gainfully.

However, this could also lead to other effects. Once a level gets freed from control by repetitive signals that come from below it, then it could start to "send signals down" to instruct those lower levels to try to detect other, different kinds of evidence. For example, during your railroad trip, perhaps you first heard those clacks on the tracks as forming a pattern of *clack-clack-clack-clack*s—that is, of beats in 4:4 time. Then you stopped hearing them at all—but then you may have suddenly switched to hearing groups of *clack-clack-clack*s—that is, of beats in 3:4 time. What made you change your representation? Perhaps some higher level just switched to forming a different hypothesis.

Also, when repetitive signals anesthetize some parts of your brain, this could release some other resources to think in new, unusual ways. This could be why some types of meditation can thrive on repetitive mantras and chants. It also could contribute to what makes some music so popular: by depriving the listener of some usual inputs, that repetitiousness could free higher-level systems to pursue their own ideas. Then, as suggested in Chapter 5-8, they could send down some "simuli" to make some lower-level resources simulate imaginary fantasies.

## Rhythmic and Musical Differences

> "Music can move us through brief emotional states, and this can potentially teach us how to manage our feelings by giving us familiarity to transitions between the states that we know and thus gain greater confidence in handling them."
> —*Matthew McCauley*

Music (or art, or rhetoric) can divert you from your mundane concerns by evoking powerful feelings that range from delight and pleasure to sorrow and pain; these can excite your ambitions and stir you to act, or calm you down and make you relax, or even put you into a trance. To do this, those signals must suppress or enhance various sets of mental resources—but why should those kinds of stimuli have such effects on your feeling and thinking?

We all know that certain temporal patterns can lead to specific mental states; a jerky motion or crashing sound arouses a sense of panic and fear—whereas a smoothly changing phrase or touch induces affection or peacefulness.[20] Some such reactions could be wired from birth—for example, to facilitate relationships between infants and parents. For then, each party will have some control over what the other one feels, thinks, and does.

Subsequently, as we grow up, we each learn similar ways to control ourselves! We can do this by listening to music and songs, or by exploiting other external things, such as drugs, entertainment, or changes of scene. Then we also discover techniques for affecting our mental states "from inside"—for example, by "hearing" that music inside our minds. (This can have a negative side, as when people complain that they can't get certain tunes out of their heads.)

Eventually, for each of us, certain sights and sounds come to have more definite significances—as when bugles and drums depict battles and guns. However, we usually each have different ideas about what each fragment of music means—particularly when it reminds us of how we felt during some prior experience. This has led some thinkers to believe that music expresses those feelings themselves; however, most such effects are much less direct:

> G. Spencer Brown 1972: "[In musical works] the composer does not even attempt to describe the set of feelings occasioned through them, but writes down a set of commands which, if they are obeyed by the reader, can result in a reproduction, to the reader, of the composer's original experience."

Perhaps Felix Mendelssohn had something like this in mind when he said, "The meaning of music lies not in the fact that it is too vague for words, but that it is too precise for words." However, some other thinkers would disagree, as noted earlier:

> Marcel Proust 1927: "Each reader reads only what is already inside himself. A book is only a sort of optical instrument which the

writer offers to let the reader discover in himself what he would
not have found without the aid of the book."

All of this raises questions that people seem strangely reluctant to ask, such
as, *"Why do so many people like music so much, and permit it to take up so
much of their lives."*[21] In particular, we ought to ask why nursery rhymes
and lullabies occur in so many cultures and societies. In *Music, Mind, and
Meaning* (Minsky 1981) I suggested some possible reasons for this: perhaps
we use those tidy structures of notes and tunes as simplified "virtual" worlds
for refining Difference-Detectors that we can then use for condensing more
complex events (in other realms) into more orderly storylike scripts.

## Difference-Networks

Whenever you want to accomplish some goal, you will need to retrieve
some knowledge about some actions or objects that might help. But
what should you do when what you have does not exactly match what
you need? Then you'll want to find some substitute that is different—
but not too dissimilar. For example, suppose that you want to sit down,
so you look for a chair, but none is in sight. However, if there were a
bench in view, then you might regard it as suitable. What leads you to
see the bench as similar when you would not so regard a book or a lamp?
What makes us selectively notice things that are likely to be relevant?
Patrick Winston (1970) suggested doing this by organizing some bodies
of knowledge into what he called *"Difference-Networks"*—for example,
these relations among types of furniture.

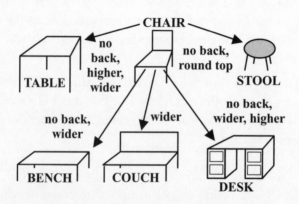

To use such a structure, one first must have some descriptions of the objects it represents. Thus, a typical concept of a chair might involve four legs, a level seat, and a vertical back, in which the legs must support the seat from below at a proper height above the floor—whereas a bench is similar (except for being wider and not having a back.

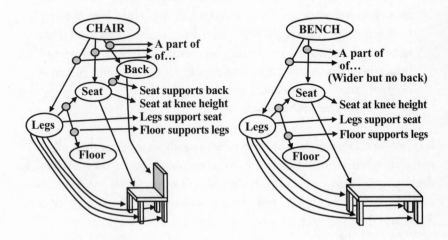

Now, when you look for a thing that matches your description of "chair," your furniture network could recognize a bench as similar. Then you can choose to accept that bench, or reject it because it is too wide or has no back.

How might we accumulate useful sets of *Difference-Links*? One way would be that, whenever we find an A that "almost works" (that is, for our present purposes) along with a B that actually works, we connect the two with a Difference-Link that represents, *"A is like B, except for a difference D."* Then such networks could also embody the knowledge we need to change what we have into what we need—as well as to suggest alternative views whenever the present one fails. Thus, such Difference-Networks could help us to retrieve memories that are relevant.

Most traditional programs were designed to use more hierarchical schemes—such as regarding a chair as an instance of *"furniture,"* and a table as just another instance. Such hierarchical classifications often help to find suitably similar things, but cannot make enough kinds of distinctions. I suspect that people use both techniques but that the "sideways" connections in our Difference-Networks are more vital to how we construct the analogies that are among our most useful ways to think about things.

## 6-5  Making Decisions

"This river which hid itself doubtless emerged again at some distant spot. Why should I not build a raft and trust myself to its swiftly flowing waters? If I perished, I should be no worse off than now, for death stared me in the face, while there was always the possibility that . . . I might find myself safe and sound in some desirable land. I decided at any rate to risk it."
—The Arabian Nights[22]

It is easy to choose among options when one appears better than all of the rest. But when you find things hard to compare, then you may have to deliberate. One way to do this would be to imagine how one might react to each possible outcome, and then, somehow, to compare those reactions, and then to select the one that seems best.

"Sensitive imagination is found in every animal, but deliberative imagination only in those that can calculate: for whether this or that shall be enacted is already a task requiring calculation."
—*Aristotle, in* On the Soul

One way a person could "calculate" would be to assign a numerical score to each choice, and then to select the largest one.

Citizen: Lately, I have been trying to choose between a country home and an apartment in town. The first one offers more spacious rooms and looks out on a beautiful mountain view. The other is closer to where I work, is in a friendlier neighborhood, but has a higher annual cost. But how could one measure, or even compare, situations that differ in so many ways?

It would be convenient if everyone could agree on the relative values of all our things. However, we each have different sets of goals and, frequently, those goals will conflict. Still, you could try to imagine how each of those situations would help or hinder you to accomplish your various goals.

> Citizen: That might just make the problem worse, because then you have to measure your feelings about the values of those various goals.

> Benjamin Franklin 1772: "When these difficult cases occur, they are difficult chiefly because while we have them under consideration all the reasons pro and con are not present to the mind at the same time; but sometimes one set present themselves and at other times another, the first being out of sight. Hence the various purposes or inclinations that alternatively prevail, and the uncertainty that perplexes us."

However, Franklin went on to suggest a way to eliminate much of that measuring:

> "To get over this, my way is, to divide half a sheet of paper by a line into two columns, writing over the one pro, and over the other con. Then during three or four days consideration I put down under the different heads short hints of the different motives that at different times occur to me for or against the measure. When I have thus got them all together in one view, I endeavor to estimate their respective weights; and where I find two, one on each side, that seem equal I strike them out: if I find a reason pro equal to some two reasons con, I strike out the three. If I judge some two reasons con equal to some three reasons pro I strike out the five; and thus proceeding I find at length where the balance lies; and if after a day or two of further consideration nothing new of importance occurs on either side, I come to a determination accordingly. And tho' the weight of reasons cannot be taken with the precision of algebraic quantities, yet when each is considered separately and comparatively and the whole lies before me, I think I can judge better, and am less likely to take a rash step; and in fact I have found great advantage from this kind of equation, in what might be called 'Moral' or 'Prudential Algebra.'"

Of course, if such a process were to conclude that several options seem equally good, then you would have to switch to another technique. You sometimes do this reflectively, but at other times the rest of your mind does this without your knowing how the decision was made. At such times you may say things like, *"I used my 'gut feelings'"* or *"I used 'intuition,'"* or you may claim that you did that *"instinctively."*

> Paul Thagard 2001: "Many persons trust their 'gut feelings' more. . . . You may have a strongly positive gut feeling toward the more interesting subject along with a strongly negative feeling about the more career-oriented one, or your feelings may be just the opposite. More likely is that you feel positive feelings toward both alternatives, along with accompanying anxiety caused by your inability to see a clearly preferable option. In the end, intuitive decision makers choose an option based on what their emotional reactions tell them is preferable."[23]

However, using the word *emotional* does not help us to see what is happening—because how "positive" or "negative" a feeling seems will still depend on how one's mental processes deal with *"all the reasons pro and con"* that Franklin addressed in that letter. Indeed, we frequently have the experience that, shortly after we make a decision, we find that it *"just does not feel right"*—and go back to reconsidering.

> Citizen: Even when options seem equally good, I still can decide. How could your kind of theory explain our peculiarly human "freedom of choice"?

It seems to me that when people say, *"I used my free will to make that decision,"* this is roughly the same as saying, *"some process stopped my deliberations and made me adopt what seemed best at that moment."* In other words, "free will" is not a process we use to make a decision, but one that we use to stop other processes! We may think of it in positive terms, but perhaps it also serves to suppress the sense that we are being forced to make a choice—if not by pressures from outside, then by causes that come from inside our own minds. To say that *"My decision was free"* is almost the same thing as to say, *"I don't want to know what decided me."*[24]

## 6-6 **Reasoning by Analogy**

*"If I had eight hours to chop down a tree, I'd spend six sharpening my axe."*
                                                    —*Abraham Lincoln*

The best way to solve a problem is to already know a solution for it—and this is why commonsense knowledge is useful. But what if the problem is one you have never seen before? How can you continue to work when you lack some of the knowledge you need? The obvious answer: you just have to guess—but how does one know how to make a good guess? We usually do this so fluently that we have almost no sense of how we are doing it, and, if someone asks about that, we tend to attribute it to mysterious traits with names like *intuition, insight, creativity,* or *intelligence.*

More generally, whenever anything attracts your attention—be it an object, idea, or a problem—you are likely to ask yourself what that thing is, why is it there, and whether it should be a cause for alarm. But as we said in Section 6-3, we can't usually say what anything *is:* we can only describe what something is *like,* and then start to think about questions like these:

> What sorts of things is this similar to?
> Have I seen anything like it before?
> What else does it remind me of?

This kind of thinking is important because it helps us to deal with new situations—and in fact, that is almost always the case, because no two situations are ever the same, and this means that we're always making analogies. For example, if the problem that you are facing now reminds you of one that you solved in the past, then you may be able to use that knowledge to solve your problem by using a procedure like this:

> *The problem that I am working on reminds me of a similar one that I solved in the past, but the method that was successful then does not quite work on the problem that I am facing now. However, if I can describe the differences between that old problem and this new one, those differences might help me to change that old method so that it will work for me now.*

We call this "reasoning by analogy" and I'll argue that this is our most usual way to deal with problems. We do this because, in general, old methods rarely work perfectly, as new situations are never quite the same. So instead, we use analogies. But why do analogies work so well? Here is the best way I've seen to explain why this is:

> Douglas Lenat 1997: "Analogy works because there is a lot of common causality in the world, common causes which lead to an overlap between two systems, between two phenomena or whatever. We, as human beings, can only observe a tiny bit of that overlap; a tiny bit of what is going on at this level of the world. . . . [So] whenever we find an overlap at this level, it is worth seeing if in fact there are additional overlap features, even though we do not understand the cause or causality behind it.

So now let's inspect an example of this.

## A Geometric Analogy Program

Everyone has heard about great improvements in computer speed and capacity. It is not so widely known that, in other respects, computers have not really changed very much in their basic capabilities. Designed originally for doing high-speed arithmetic, it was usually assumed that this was all computers would ever accomplish—which is why they were misnamed "computers."

However, people soon began to write programs to deal with nonnumerical things such as linguistic expressions, graphical pictures, and various forms of reasoning. Also, instead of following rigid procedures, some of those programs were designed to search through wide ranges of different attempts—so that they could solve some problems by trial and error, instead of by using pre-programmed steps. Some of these early nonnumerical programs became masters at solving some puzzles and games, and some were quite proficient at designing new kinds of devices and circuits.[25]

Yet despite those impressive performances, it was clear that each of these early "expert" problem-solving programs could operate only in some narrow domain. Many observers concluded that this came from some limitation of the computer itself. They said that computers could solve only "well-defined problems" and would never be able to cope with ambiguities, or to use the kinds of analogies that make human thinking so versatile.

To make an analogy between two things is to find ways in which they are similar—but when and how do we see two things as similar? Let's assume that they share some common features but also have some differences. Then how similar they may seem to be will depend upon which differences one decides to ignore. But the importance of each difference depends upon one's current intentions and goals. For example, one's concern with the shape, size, weight, or cost of a thing depends on what one plans to use it for—so, the kinds of analogies that people will use must depend upon their current purposes. But before the Difference-Engine idea was conceived, few people believed that machines could ever have goals or purposes.

> Citizen: But if your theory of how people think depends on using analogies, how could any machine do such things? People have always told me that machines can only do logical things, or solve problems that are precisely defined—so they cannot deal with hazy analogies.

To refute such beliefs, Thomas G. Evans (1963) wrote a program that performed surprisingly well in what many people would agree to be ambiguous, ill-defined situations. Specifically, it answered the kinds of questions in a widely used "intelligence test" that asked about "geometric analogies." For example, a person was shown the picture below and asked to choose an answer to: "*A is to* B *as* C *is to which of the other five figures?*" Most older persons choose figure 3—and so did Evans's program, whose score on such tests was about the same as that of a typical sixteen-year-old.

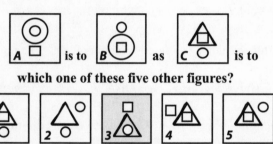

**which one of these five other figures?**

In those days, many thinkers found it hard to imagine how any computer could solve such problems, because they felt that choosing an answer must come from some "intuitive" sense that could not be embodied in logical rules. Nevertheless, Evans found a way to convert this to a far less mysterious kind of problem. We cannot describe here all the details of his program, so we will show only how its methods resemble what people do in such situations. For if you ask someone why they chose figure 3, they usually give an answer like this:

> You can change *A* to *B* by moving the big circle down, and you can change *C* to *3* by moving the big triangle.

This statement expects the listener to understand that both clauses describe something in common—even though there is no big circle in figure 3. However, a more articulate person might say:

> You can change *A* to *B* by moving the largest figure down, and you can change *C* to *3* by moving the largest figure down.

Now those two clauses are identical—and this suggests that one could use a three-step process based on these kinds of descriptions. First, invent descriptions for each of top row of figures. For example, these might be

> Figure *A* shows *high large, high small, and low small objects.*
> Figure *B* shows *low large, high small, and low small objects.*
> Figure *C* shows *high large, high small, and low small objects.*

Next, invent an explanation for how *A* might have been changed to *B*. For example, this might simply be:

> *Change "high large" to "low large."*

Finally, use this to change the description of figure *C*. The result will be

> Figure *C* shows *low large, high small, and low small objects.*

If this prediction—of how figure C has been changed—matches one of the possible answers more closely than any other one, that's what we'll choose for our answer! In fact, it matches only figure 3—which is the one that most people select. (If it matches two or more figures, Evans's program starts over again by making different descriptions of the same pictures.) The program performed as well on such tests as did typical fifteen-year-olds.

Of course, whenever we need to make a choice, the differences that will concern us most will depend on our goals. If Carol wants merely to build an arch, then all of the forms in the figure below may seem adequate—but if she plans to put more objects on top of her arch, then the one on the right will be less suitable.

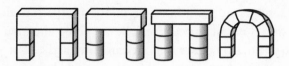

Although these particular "geometric analogy" problems are not very common in everyday life, Evans's program shows the value of being able to change our descriptions until we find ways to describe different things so that they seem more similar. This often enables us to use our knowledge about one kind of thing to understand some other, different kind of thing—and discovering *new ways to look at things* is one of our most powerful commonsense processes.

> George Pólya 1954: "We can learn the use of such fundamental mental operations as generalization, specialization, and the perception of analogies. There is perhaps no discovery, either in elementary or in an advanced mathematics, or, for that matter, in any other subject, that could do without these operations, especially without analogy."

Notice that to make and use an analogy, one must work on three different levels at once: *(1) descriptions of the original objects, (2) descriptions of their relationships,* and *(3) descriptions of the differences between those relationships.* Of course, as we saw in Chapter 5-2 and 5-3, none of these descriptions should be too concrete (or they won't apply to other examples), and none of them should too abstract (or they won't be able to represent the differences that are relevant).[26]

## 6-7  Positive vs. Negative Expertise

"Never interrupt your enemy when he is making a mistake."
        —*Napoleon Bonaparte*

"I have learned throughout my life as a composer chiefly through
my mistakes and pursuits of false assumptions, not by my expo-
sure to founts of wisdom and knowledge."
        —*Igor Stravinsky*

At the very beginning of Chapter 1, we pointed out that many feelings we
think of as "positive" are partly based on censoring aspects of things that we
might otherwise think of as negative. Thus a certain situation could seem
"pleasant" to mental processes that are currently active, but might seem
quite unpleasant to other processes that are currently being suppressed.

For example, the process of raising a human child requires years of
work and worry to feed, clean, dress, teach, shelter, and protect that
child. What kind of incentive could make one forgo so many other goals
and become so selfless and other-directed? Of course, we see mother-
love as positive—but if people had not evolved ways to suppress so many
kinds of discouraging prospects, no one would have had any descendants. Here are a few more examples in which we conceal disagreeable
aspects of things:

Humor: Humor is usually seen as positive, despite the fact that
most jokes are basically negative—in the sense that they almost
always speak about things that a person should not do, because
they are socially prohibited or simply absurd or ridiculous. (See
Minsky 1980.)

Decisiveness: We often speak of "making a choice," as though
this were a deliberate act. However, that "action" may, in fact,
be nothing more than the moment at which you *stopped* some
process that was comparing alternatives—and then, by default,
you simply adopted the one that was then at the top of some list.
In such a case, a person may speak of using "free will"—but an
observer could also see it as nothing more than a sort of admission
(or even a boast) that one does not have a clear idea about what
mental process produced that result.

Beauty: We tend to see beauty as positive—but when people say something is beautiful, and you ask what makes them like it so much, they are likely to act as though under attack, or to simply insist, "I just like it." This might suggest that some process is working to keep them from noticing defects or blemishes.

Pleasure: When we think that we're choosing the option that pleases us most, the selection may actually come from some process that has silenced all its competitors. This, as every addict knows, makes it hard to wish for anything else. If so, then the more pleasure we feel, the more negative may be that hidden effect on the rest of our mental processes! In such cases, *"I am enjoying this,"* could mean, *"I want to remain in my present state, so I'll try to suppress whatever might change it."*

We also can sometimes disable a process without directly suppressing it, by arousing one that competes with it. For example, Chapter 8-3 showed how one can hold off sleep by imagining a disturbing situation. Or one can simply repeat a stimulus until there is no further response to it, as in the old tale *The Boy Who Cried Wolf.*

Teacher: I was taught that learning works primarily by using pleasure to "reinforce" those connections that have led to success—whereas failures deter and discourage us. Therefore, teachers should make each lesson pleasant by giving rewards and encouragements.

This idea that each learning experience should be "positive" was largely based on results from research that was mainly done with pigeons and rats. Then many educators generalized this to human students, concluding that it would be best *to teach every subject by very small steps, so that pupils will usually meet with success.* However, to understand a complex situation, one also needs to learn how things can go wrong so that one can evade the most common mistakes.

Teacher: Surely we can develop ways to teach people how to avoid mistakes in ways that are pleasant and positive. We can reward them for getting through complex tasks, for detecting and reacting to mistakes, and for persistence and originality. Is there any basic reason why resourcefulness, itself, cannot be reinforceable?

In Chapter 9-6, I'll argue that the answer is both yes and no, because it involves what might seem a paradox: it is "pleasant" to have accomplished a difficult task, but this almost always involves some transient episodes of severe distress and discomfort. So for students to learn to do such things, I will argue in Chapter 9-4 that those students will need to develop ways to tolerate—and even to enjoy—those painful kinds of episodes. Also, on the other side, here are a few other reasons why only rewarding successful attempts may not be a very good strategy:

*Reinforcement can lead to rigidity.* If a system already works, additional "reinforcement" could make some internal connections stronger than they need to be. Then this could make it harder for that system to later adapt to new kinds of situations.

*Reinforcement can have bad side effects.* If a certain resource has worked so well that other processes have come to depend on it, then any change you make in that resource will be likely to damage the performances of those other processes (because unplanned changes usually make things worse).

*Papert's Principle:*[27] Some of the most crucial steps in our mental growth are based not simply on acquiring new skills, but on developing better high-level resources to help us select which already existing skills to use.

I certainly do not mean to suggest that positive reinforcement is bad—but we often learn more from a failure than from a success, especially when we need to learn not only which methods are likely to fail, *but also how and why those failures occur,* as well as *what might have caused our thoughts to go wrong.* In other words, one learns much more when one investigates, rather than merely celebrates.

A skeptical teacher might wonder if the ideas in this section have been confirmed by any animal experiments. I would have to answer no, because most of the processes that we've discussed could take place only in reflective levels that no other animals possess.

Student: I don't see why reflection has to be involved. Why could we not learn from the times we fail, simply by breaking the connections we used—so that after you've made a bad mistake, your brain will tend less to do it again?

Removing connections might sometimes work, but exposes us to a different risk: whenever you make a change in a system's connections, this is likely to impair those other resources that also depend on those same connections. If you don't understand how a system works, then you're in danger of making it worse by blindly correcting apparent mistakes.

> Programmer: Every attempt to improve a program is likely to introduce new bugs. That's why new programs so often contain very big sections of ancient code: no one remembers quite how they work, and hence they're afraid to change them. So, if something is wrong that you need to fix, then it is best to install a small, local, "patch" and hope that the rest of the system will still work.

More generally, you can usually start to improve a skill by experimenting with many small steps—*but eventually no more such changes may help, because you have reached a local peak.* Then further improvement may require you to endure some discomfort and discouragement. Here is a simple example of this:

> Charles happens to be in Tanzania. He wants to be at the highest possible altitude while still remaining on solid ground. So at every point he takes his next step in the direction of steepest ascent. Eventually, he may end up at the top of a very small hill, but if he is lucky he'll finally get to the top of Mount Kilimanjaro. However, his strategy will never reach the summit of Mount Everest—because every such route includes some downward steps.

Of course, this also applies whenever we try to improve a mental ability. For a time, we can use that method of "steepest ascent" by making many small, pleasant changes. But then, to make any further gain, we'll have to endure at least some distress. So, while pleasure helps us learn easy things, we must learn to "enjoy" some suffering when it comes to learning things that need larger-scale changes in how we think. Accordingly, Chapter 9-4 will suggest that trying to make education too pleasant might hinder children from learning to climb the conceptual mountains inside their minds.

*  *  *

This chapter discussed the bodies of commonsense knowledge that human beings need to get along in the civilized world. I touched on many questions about what we mean by common sense, what goals are and how they work, how we usually reason by making analogies, and how we might guess which items of knowledge might be relevant to how we make our decisions. I also emphasized the role of "negative" knowledge about how to avoid making common mistakes.

However, it is not enough just to know a lot; we also must put that knowledge to work. So the following chapter will talk about the processes that we employ in our many everyday ways to think.

# 7
# THINKING

Which feature most clearly distinguishes us from the rest of our animal relatives? Surely our most outstanding such trait is our knack for inventing new Ways to Think.

> Romanticist: You suggest that our finest distinction is thinking, yet perhaps the richness of our mental experience is even more special—as in our sense of being alive, or the joy of turning our intellect off to enjoy a sunset or listen to birds, or to perform a spontaneous song or dance.

> Determinist: People use words like *spontaneous* to make themselves feel that they aren't constrained. But perhaps that sense of enjoying ourselves is merely a trick that some parts of our brains use to make us do what they want us to do.

I doubt that we ever stop thinking, because that word refers, at different times, to a huge range of intricate processes—and many of these are hidden from us.

> Citizen: If our everyday thinking is so complex, then why does it seem so straightforward to us? If its machinery is so intricate, how could we be unaware of this?

That illusion of simplicity comes from forgetting our infancies, in which we grew those abilities. As children, we learned how to pick up blocks and arrange them into rows and stacks. Then as each new group of skills matured, we built yet more resources on top—just as we learned to plan and build more elaborate arches and towers. So each of us built, in those early times, the towers of aptitudes that we call "minds."

But now, as grown-ups, it seems to us that *we have always been able to reason and think*—because we learned those skills so long ago that we cannot recall having learned them at all. It took us many years of hard work to develop our more mature Ways to Think, but whatever records remain of this have somehow become inaccessible. What could have made us all victims of that "amnesia of infancy"? I don't think this is simply because we "forgot." Instead, I suspect that it results from our developing new and better ways to represent both physical and mental events—and some of these methods became so effective that they replaced our previous ones. Now, if those old records still exist, we can no longer make sense of them.

In any case, all of us find it so easy to think that we rarely ask good questions about what thinking is and how it might work. In particular, we like to celebrate grand accomplishments in the sciences, arts, and humanities—but we scarcely ever recognize the marvels of everyday commonsense thinking. Indeed, we often see thinking as more or less passive, as though our ideas just *"occur to us"* and we don't deserve any credit for them—as when we say, *"A thought just entered my mind"* instead of *"I just constructed a good new idea."* Similarly, we scarcely ever wonder about what chooses which subjects we think about.

> One of the wooden doors in my home bears scratches made more than a decade ago. Our dog Jenny is gone, but the scratches remain. I notice them only a few times a year, though I pass by that door several times every day.

In every hour of every day you encounter great numbers of things and events, yet only a few of them "get your attention" enough to raise such questions as, *"What is that object and why is it there,"* or *"Who or what caused that happening?"* Thus much of the time, your thoughts seem to proceed in a smooth, steady flow in which you scarcely ever reflect on how you got from each step to the next.

At other times, though, your mind seems to wander without any sense of aim or direction: first you dwell on some social affair; then you reflect on some past event; next you're beset by a hunger pang, or the thought of a payment that's overdue, or an impulse to fix that faucet drip, or an urge to tell Charles how you feel about Joan. Each item reminds you of something else until one of your mental Critics interrupts with, *"This isn't getting you anywhere,"* or *"You must try to get more organized."*

However, this chapter will be mainly concerned with what happens when your thinking is aimed toward some definite goal but then encounters an obstacle, as when you say to yourself, *"I can't pack all this into this box—and besides, that would make it too heavy to lift."* Such a mental event is likely to interrupt most of your current processes, and make you stop to deliberate: *"It looks like this will take several trips, but I don't want to spend that much time on this."* At such a point, your efforts may switch from the goal of packing that box to higher-level reflections about selecting a different subject to think about.

This chapter will focus on the idea that each person has many different ways to think—but first, perhaps, we ought to ask why we have so many of these. One answer is that our ancestors evolved through hundreds of different kinds of environments, each of which required ways to deal with new kinds of situations. However, we never discovered a single, uniform scheme that could cope with all those different conditions. Consequently, over eons of time, our brains evolved a good many different ways to avoid the most common kinds of mistakes.

This suggests another reason why we evolved so many different Ways to Think: if your thoughts were controlled in only a single way, you would be in danger of becoming a monomaniac. Of course, such accidents have constantly happened in the course of human history, but the genes of those individuals do not usually propagate because their bearers lack versatility. Indeed, as we noted in Chapter 6-2, although evolution is often described as a process of selecting beneficial changes, the bulk of evolution's work involves rejecting changes that have bad effects. The result is that most species evolve at the edge of some narrow zone between the safeties they know and the dangers they don't.

Psychiatrist: That safety zone can be narrow, indeed. Most of the time, most minds function well, but sometimes they get into various states in which they can scarcely function at all—and then we say that they're mentally ill.

Physiologist: Surely most such disorders have medical causes—such as traumatic injuries, chemical imbalances, or diseases that damage our synapses.

Programmer: Perhaps, but we should not assume that *all* such disorders have nonmental causes. When a software virus infects a computer and changes some data on which the programs depend, the hardware may not be damaged at all, but the system's behavior may totally change.

Similarly, a new destructive goal or idea—or a change in one's Critics or Ways to Think—could gain control of so much of a person's resources and time that we would seem to be seeing a different mind.

Sociologist: Perhaps it's the same for a social organization, when the policies of a sect or cult include ways to recognize those recruits in whom its ideas and beliefs will propagate.

In any case, as we evolved the machinery that could support new Ways to Think, this also forced us to evolve ways to learn which of those strategies would be useful for dealing with new classes of situations or problem-types.

## 7-1  What Selects the Subjects We Think About?

What selects what we'll think about next, from among all our various interests—and how long will we persist with each? Let's consider a typical everyday incident:

*Joan needs to write a project report but has not made much progress on it. Discouraged, she sets those thoughts aside and finds herself roaming about her house with no particular goal. She passes an untidy stack of books and stops for a moment to straighten them out. But then she "gets" a new idea, so she goes to her desk to type a note. She starts to type—but finds that the T on her keyboard is stuck. She knows how to fix this, but she worries that then she might forget that new idea—so instead she makes a handwritten note.*

What led Joan to notice that pile of books? Why did that that idea "occur" to her now instead of at some other time? Let's look more closely at these events.

> **Joan has not made much progress.**  Some mental Critic must have noticed this and suggested that she "take a break."
>
> **Discouraged, she sets those thoughts aside.**  How will Joan later get back to her previous states? Section 7-6 will talk about how she could make—and later retrieve—the contexts of some of her previous thoughts.
>
> **Joan is roaming without any goal.**  Or so it may seem—but most animals have instincts to maintain their "territories" or nests. Usually, Joan walks right past this spot, but right now she is "making rounds" because of being mainly controlled by Critics that aim to maintain the tidiness of her home.
>
> **She passes an untidy stack of books and stops for a moment to straighten it out.**  Why doesn't Joan stop to read those books instead of just trying to tidy them up? This is because, right now, she represents those books as untidy objects rather than as containers of knowledge.
>
> **But then she "gets" a new idea.**  When people say, "It occurred to me," this shows how limited is the extent to which we can reflect on how we produce our ideas.
>
> **Joan goes to her desk to type a note.**  Here, Joan is using a model of herself that represents her knowledge about the qualities of her short-term memories. She knows that when she "gets" an idea, she cannot depend on remembering it—and so she puts her housekeeping on hold to make a more permanent record.

Perhaps most of the time we mainly react to external events without much sense of making decisions. However, our higher-level thinking depends much more on our wishes, fears, and larger-scale plans—and this raises many questions about how we spend our mental time:

What schedules our large-scale plans?
What reminds us of things that we promised to do?

How do we choose among conflicting goals?

What decides when we should quit or persist?

So long as everything goes well, such questions rarely occur to you, and your thoughts proceed in a steady, smooth flow. Each minor obstacle makes only small changes in how you think, and if you "notice" these at all, they merely appear as transient feelings or as fleeting ideas. But, when more serious obstacles persist and keep you from making progress, then various Critics intervene to make larger changes in how you think.

## 7-2 The Critic-Selector Model of Mind

"I have yet to see any problem, however complicated, which, when you looked at it in the right way, did not become still more complicated."

—*Poul Anderson*

We frequently change what we're thinking about, without noticing that we are doing this—because it is mainly when some trouble comes that we start to reflect about thinking itself. Thus, we don't recognize a problem as "hard" until we've spent some time on it without making any significant progress. Even then, if that problem does not seem important, you might just abandon that line of thought and simply turn to some other subject.

However, if you have an important goal, then it is useful to notice that you are stuck—and it will be even more useful if you also can recognize that you're being blocked by a certain particular type of barrier, obstacle, impasse, or snag. For if you can diagnose the particular Type of Problem you face, then you can use that knowledge to switch to a more appropriate Way to Think.

This suggests a model of mind based on reacting to "cognitive obstacles." We'll call this the "*Critic-Selector.*"

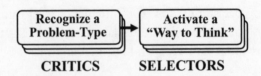

THE CRITIC-SELECTOR MODEL OF MIND

Each Critic in this diagram can recognize a certain species of "Problem Type." When a Critic sees enough evidence that you now are facing its type of problem, then that Critic will activate what we shall call a *"Selector,"* which tries to start up a set of resources that it has learned is likely to act as a Way to Think that may help in this situation.

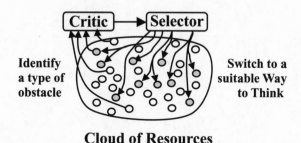

**Cloud of Resources**

The simplest Critic-Selector system could consist of little more than a collection of *If→Do* rules like these:

*If* a problem seems familiar, *Use* reasoning by analogy.
*If* it seems unfamiliar, *Change* the way you're describing it.
*If* it seems too difficult, *Divide* it into several parts.
*If* it still seems difficult, *Replace* it by a simpler problem.
*If* none of these work, *Ask* someone for help!

Student: I don't need to insert those Selectors. Why not design each Critic to directly turn on a set of resources for solving the problem it recognizes—just as you showed in Chapter 1-5?

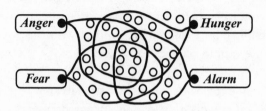

I suspect that it was extremely difficult for our ancestors to invent useful new Ways to Think—until they evolved ways to make new Selectors by combining smaller sets of parts. So although that student is partly right, I suspect that our human brains evolved to include more such machinery rather than less. For example, each Critic could recommend the use of not just one but several different Selectors:

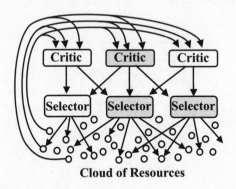

**Cloud of Resources**

In any case, it seems safe to assume that the brains of our early ancestors did not include Selectors at first. But as we came to face increasingly complex problems, it would have become increasingly harder to invent new Critics that worked efficiently—and it is always easier to invent new, useful structures when one already has stocks of older parts that one can adapt and recombine. (For example, the Wright brothers made their first airplane work by using motorcycle parts.) More generally, until one can describe a process in terms of the functions of smaller parts, it will be difficult to reflect on how that process works—and then to invent useful variations of it.

For example, Section 8-4 will describe some structures called *"K-lines"* that our brains could use to construct new mental objects and processes by combining parts of older ones.

Of course, when one activates two or more Critics or Selectors, this is likely to cause some conflicts, because two different resources might try to turn a third resource both *on* and *off.* To deal with this, we could design the system to use various policies like these:

Choose the resource with the highest priority.
Choose the one that is most strongly aroused.
Choose the one that gives the most specific advice.
Have them all compete in some "marketplace."

However, while competitive strategies might suffice for fairly simple brains, I doubt that such schemes would work on larger scales, unless supervised by processes that use additional knowledge about how to settle particular kinds of conflicts. For example, one might use these more reflective policies:

> *If* too many Critics are aroused, *then* describe the problem in more detail.
> *If* too few Critics are aroused, *then* make the description more abstract.
> *If* important resources conflict, *then* try to discover a cause for this.
> *If* there has been a series of failures, *then* switch to a different set of Critics.

A brain with good memories of the recent past could recognize later, after the fact, that certain selections had serious flaws—and then could proceed to try to find ways to refine the Critics that made those mistakes.

> After I selected that method, I realized that I knew a better one.
> I now see that the action I took had an irreversible side effect.
> I treated something as an obstacle, but now see that it was valuable.
> That method did not actually work, but I learned a lot from using it.

However, to recognize those kinds of events, we would need more "reflective" Critics that work at higher levels—and this suggests that our model of mind should include Selectors and Critics at every level.[1]

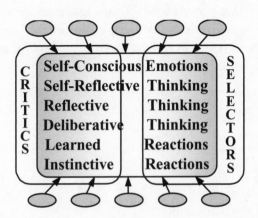

## 7-3 Emotional Thinking

"There is a very fine line between 'hobby' and 'mental illness.'"
—*Dave Barry*

Most of the time, your thinking proceeds in uneventful streams—but when you run into obstacles and none of your usual methods help, you may resort to mental strategies that may appear to be somewhat irrational. For example, when tempted to abandon your task, you can renew your motivation by bribing yourself with imagined rewards, or with threats of the prospect of failure—or you could try to shame yourself by imagining how you (or your Imprimers) might feel if your performance conflicted with your highest values. For even a brief flash of impatience, or of anger or desperation, can cut through what seems like a hopelessly tangled knot.

Each such "emotional" Way to Think can lead to different ways to deal with things—either by making you see things from new points of view or by increasing your courage or doggedness. If this initiates a large-scale cascade—and if those changes last for long enough, then you (or your friends) may recognize this as a change in your emotional state.

How long do those states of mind persist? Some last for no more than the blink of an eye, but infatuations persist for days or weeks. However, when "dispositions" endure for weeks or years, we regard them to be aspects of a person's personality, and call them "characteristics" or "traits."

For example, when solving a problem, some people tend to accept solutions that still have some deficiencies—so long as those answers work well enough. You might describe such persons as "realistic," "pragmatic," or "practical." Other persons tend to insist that every potential flaw must be fixed—and you might call such people "fastidious," except when they make you uncomfortable, in which case you call them "obsessive" instead. Other such dispositions include being

Cautious vs. Reckless
Unfriendly vs. Amicable
Visionary vs. Practical
Inattentive vs. Vigilant
Reclusive vs. Sociable
Courageous vs. Cowardly

In the course of everyday thought, a person is likely to switch among such attitudes, often without any notice of this. However, when we encounter more serious trouble, our Critics may make enough changes to start the large-scale cascades that we describe in terms of emotional states.

> Psychiatrist: What would happen if too many Critics were active? Then your emotions would keep changing too quickly. And if those Critics stopped working at all, then you'd get stuck in just one of those states.

Perhaps we can see an example of this in Antonio R. Damasio's book *Descartes' Error*, which describes a patient named Elliot, who had lost some parts of his frontal lobes in the course of having a tumor removed. After that treatment, he still seemed intelligent, but his friends and employers had the sense that Elliot was "no longer himself." For example, if asked to sort some documents, he was likely to spend an entire day at carefully reading just one of those papers—or at trying to decide whether to classify them by name, or by subject or size or date or weight.

> Damasio 1995: "One might say that the particular step of the task at which Elliot balked was actually being carried out too well, and at the expense of the overall purpose. . . . True, he was still physically capable and most of his mental capacities were intact. But his ability to reach decisions was impaired, as was his ability to make an effective plan for the hours ahead of him, let alone to plan for the months and years of his future."

The damaged parts of Elliot's brain included certain connections (to the amygdala) that are widely believed to be involved with how we control our emotions.

> Damasio 1995: "At first glance, there was nothing out of the ordinary about Elliot's emotions. . . . However, something was missing. . . . He was not inhibiting the expression of internal emotional resonance or hushing inner turmoil. He simply did not have any turmoil to hush. . . . I never saw a tinge of emotion in my many hours of conversation with him: no sadness, no impatience, and no frustration with my incessant and repetitious questioning."

This led Damasio to suggest that "reduced emotion and feeling might play a role in Elliot's decision-making failures." However, I'm inclined to turn that around to suggest that it was Elliot's *new inability to make such decisions that reduced his range of emotions and feelings.* For perhaps what got damaged in Elliot's brain were mainly the Critics (or their output connections) that formerly set off the processes that we recognize as emotional states. For now he has lost those precious cascades—and, hence, the emotions that he once displayed—because he no longer can use those Critics to choose which emotional states to use.

This still leaves us with many questions about how such systems might be organized. So now this chapter will try to describe some of our many Ways to Think, as well as some of the Critics we use to diagnose the types of problems that we frequently face.

## 7-4  What Are Some Useful Ways to Think?

"When you want people to think you are brilliant, just imagine the worst thing you could do and then do precisely the opposite."
—*Naomi Judd*

It ought to be a central goal—both for AI and for Psychology—to find some systematic means to classify the ways we try to overcome different types of obstacles. But because no good such scheme has yet appeared, I'll just list some examples of Ways to Think, beginning with these two extremes:

> *Knowing How.*  The best way to solve a problem is to already know a way to solve it. However, we may not be able to retrieve that knowledge—and often we don't even know that we have it.
> *Searching Extensively.*  When one has no better alternative, one could try to search through all possible chains of actions. But that method is not often practical because such searches grow exponentially.

However, we each know many other Ways to Think that lie between those two extremes, and help to make those searches more feasible.

> *Reasoning by Analogy.*  When a problem reminds you of one that you solved in the past, you may be able to adapt that case to the

present situation—if you have good ways to tell which similarities are most relevant.

*Dividing and Conquering.* If you can't solve a problem all at once, then break it down into smaller parts. For example, every difference we recognize may suggest a separate subproblem to solve.

*Reformulating.* Find a different representation that highlights more relevant information. We often do this by making a verbal description—and then "understanding" it in some different way!

*Planning.* Consider the set of subgoals you want to achieve and examine how they affect each other. Then, with those constraints in mind, propose an efficient sequence for achieving them.

We all know other techniques that work by first solving a different problem.

*Simplifying.* Often a good way to solve a difficult problem is first to solve a simpler version that ignores some features of that problem. Then any such solution may serve as a sequence of stepping-stones for solving the original problem.

*Elevating.* If you are bogged down in too many details, describe the situation in more general terms. But if your description seems too vague, switch to one that is more concrete.

*Changing the subject.* Whatever you are working on now, if you get discouraged enough, you can always abandon it and simply switch to a different task.

Here are some more reflective Ways to Think:

*Wishful thinking.* Imagine having unlimited time and all the resources that you might want. If you still can't envision solving the problem, then you should reformulate it.

*Self-reflection.* Instead of further pursuing a problem, ask what makes that problem seem hard, or what you might be doing wrong. This can suggest some better techniques—or, instead, better ways to spend your time.

*Impersonation.* When your own ideas seem inadequate, imagine someone better at this, and try to do what that person would do. Myself, I do this frequently, by imitating Imprimers and teachers.

We also use many other Ways to Think.

> *Logical Contradiction.* Try to prove that your problem cannot be
> solved, and then look for a flaw in that argument.
> *Logical Reasoning.* We often try to make chains of deductions.
> However, this can lead to wrong conclusions when our assump-
> tions turn out to be unsound.[2]
> *External Representations.* If you find that you're losing track of
> details, you can resort to keeping records and notes, or drawing
> suitable diagrams.
> *Imagination.* One can avoid taking physical risks if one can predict
> *"What would happen if"* by simulating possible actions inside the
> mental models that one has built.

Of course, if you are not completely alone, you can try to exploit your
social resources.

> *Cry for help.* You can behave in ways that may arouse your compan-
> ions' sympathies.
> *Ask for help.* If your status is high enough, you can persuade or
> command someone else to help—or even offer to pay them.

Thus, everyone has many ways to think, and the following section will
discuss how your Critics choose which ones to use. However, everyone
always has one "last resort"—namely, simply to give up and quit!

> *Resignation.* Whenever you find yourself totally stuck, you can shut
> down the resources you're using now and relax, lie back, drop
> out, and stop. Then the "rest of your mind" may find an alterna-
> tive—or conclude that you don't have to do this at all.

## 7-5 What Are Some Useful Types of Critics?

"Don't pay any attention to the critics. Don't even ignore them."
—*Sam Goldwyn*

We are always developing new Ways to Think, but how do we choose
which one to use? Our Critic-Selector Model of Mind assumes that our

Critics help to recognize the kinds of predicaments or prospects we face, and then recommend selections of ways in which we might deal with those situations. Our Critics must be among our most precious resources, and each person develops them in different ways—which could be partly responsible for each person's individuality.

But how could our Critics catalog all the impasses, obstacles, and snags that make some problems hard to solve? It would be an important goal, both for people and for machines, to have systematic classifications of all the kinds of problems we frequently face. However, we don't yet have orderly ways to do this[3]—so here we'll merely try to list a few types of Critics that people use.

**Innate Reactions and Built-in Alarms.** Many types of external events arouse detectors that make us quickly react—such as when an object rapidly moves toward you, or a light is too bright, or you touch something hot. We're also born equipped with ways to detect certain conditions *inside* our skins—such as abnormal blood levels of chemicals—along with built-in connections that make us react to correct those conditions without any need to think about them.

However, some unexpected touches, sights, or smells—or feelings of hunger, fatigue, or pain—*do* interrupt the flow of our thoughts. For we'd never survive through our infancies unless such emergencies (or opportunities) could pull us away from our reveries. We can sometimes succeed at suppressing a sneeze, or at stopping ourselves from scratching an itch. But it is hard to ignore a baby's cry, an insistently ringing telephone, or an amorous opportunity, and when you try to hold your breath, you cannot resist the alarm of impending asphyxia.

**Learned Reactive Critics.** A typical infant will simply cry when exposed to high levels of noise—but later a child may learn to react by moving to a quieter place. And eventually we learn to deal with obstacles by using "deliberative" thinking about them.

**Deliberative Critics.** When our first attempt fails to solve a problem, we can often discover alternatives, by thinking about what might have gone wrong. Here are some tricks we can use for this:

An action did not achieve the expected effect. (Find a better way to predict.)

Something I did had a bad side effect. (Try to undo some previous choice.)

Achieving one goal made another one harder. (Try them in the opposite order.)

I need additional information. (Search for another relationship.)

*This method works so well that I should try to use it more frequently.*

**Reflective Critics.** When you try to solve problems by trial and error, you use your critics as "diagnosticians"—either to verify that you're making progress or to suggest alternative ways to proceed. See Singh 2003b.

I've made many attempts with no success. (Select another way to think.)

I've repeated the same thing several times. (Some other process must be stuck.)

Achieving a subgoal did not attain its goal. (Split up the problem in another way.)

This conclusion needs more evidence. (Propose a better experiment.)

*This method works so well that I should apply it to other realms.*

**Self-Reflective Critics.** When you can't control the resources you need, or try to achieve too many goals at once, then you may start to criticize yourself:

I have been too indecisive. (Try a method that worked on a similar problem.)

I missed a good opportunity. (Switch to a different set of Critics.)

I yield to too many distractions. (Try to exercise more self-control.)

I don't have all the knowledge I need. (Find a good book or go back to school.)

*This works so well that I should make myself better at it.*

**Self-Conscious Critics.** Some assessments may even affect one's current image of oneself, and this can affect one's overall state:

None of my goals seem meaningful. (Depression.)

I'm losing track of what I am doing. (Confusion.)

I can achieve any goal I like! (Mania.)
I could lose my job if I fail at this. (Anxiety.)
My friends might disapprove of this. (Insecurity.)

*This works so well that I should make it my specialty!*

In Chapter 3-5 we noted that the word *Critic* often has a negative quality because it is most often applied only to persons who point out deficiencies. And indeed, it would be hard to describe our Correctors, Suppressors, and Censors without using negative terms like *inhibit, prevent,* or *terminate.* However, words like *positive* and *negative* usually do not make sense by themselves, because recognizing when something goes wrong is frequently a critical step toward success. This is why I added an "Encourager" to the end of each list above, to make room for "Positive Critics" that can assign more priority, time, or resources to the strategy that you are currently using. Furthermore, Chapter 9-4 will argue that sometimes you will need to endure some discomfort in the course of achieving a goal—and then you may need some Encouragers to make yourself persist with your plan, no matter that this involves some suffering.

## How Do We Learn New Selectors and Critics?

"Honest criticism is hard to take, particularly from a relative, a friend, an acquaintance, or a stranger."
—*Franklin P. Jones*

The first time you're faced with some difficult problem, it will take you some time to find its solution—but in the future you will find it easier to deal with other, similar situations. This must be because of what you learned from that previous experience, but what did you actually learn and how?

Perhaps the simplest way to learn from solving a problem would be merely to add a new *If →Do* rule that says "Apply the methods I recently used whenever I face a similar problem." However, if solving that problem took a long time, then one ought to ask instead, *"What kept me from solving that problem more quickly?"* For if it took a long time to find that solution, then one should try to criticize the methods one used for finding that answer. So Chapter 8-5 will argue that whenever a problem turns out to be "hard," then we should try to assign the credit for our success, not to the final act itself, but to only those parts of our thinking that actually helped to find the solution.

In other words, we can sometimes improve our Ways to Think by creating higher-level Selectors and Critics that help to reduce the sizes of the searches we make. However, to make such kinds of credit assignments, we'll need to use higher reflective levels of thinking than those that have hitherto been proposed in most traditional "theories of learning."

How do we organize our collections of Critics? How do we make them and how do we change them? Do some of our Critics scold other ones when they produce poor performances? Are certain minds more productive because their Critics are better organized?

How do we organize our collections of Ways to Think? How do we make them and how do we change them? Do some of them recognize when other ones tend to produce poor performances? Are certain minds "more intelligent" because their Ways to Think are better organized?

The following sections will argue that, today, we have no plausible answers to questions like these—and that these issues should be recognized as central to the development of psychology.

## 7-6 Emotional Embodiment

Many thinkers have maintained that emotional states are closely involved with our bodies—and that this is why we so often can recognize Happiness, Sadness, Joy, or Grief from a person's expressions, gestures, and gaits. Indeed, some psychologists have even maintained that those bodily activities do not merely "express" our emotions, but actually cause them:

> William James 1890: "Our natural way of thinking about . . . emotions is that the mental perception of some facts excites the mental affection called the emotion, and that this latter state of mind gives rise to the bodily expression. My theory, on the contrary, is that the bodily changes follow directly the perception of the exciting fact, and that our feeling of the same changes as they occur *is* the emotion."

For example, James suggests that when you sense that a rival is insulting you, this makes you clench your fist and strike—and that your anger results directly from your perceiving those physical activities. However, this makes little sense to me, because what James calls the "exciting fact" of clenching your fist cannot come first, but must come after your brain perceived that you were being insulted. Nevertheless, James argues that such intermediate thoughts could not have such strong effects by themselves:

> William James 1890: "If we fancy some strong emotion and then try to abstract from our consciousness of it all the feelings of its bodily symptoms we find we have nothing left behind, no "mind stuff" out of which the emotion can be constituted, and that a cold and neutral state of intellectual perception is all that remains. . . . [I cannot imagine] what kind of an emotion of fear would be left if the feeling neither of quickened heart beats nor of shallow breathing, neither of trembling lips nor of weakened limbs, neither of goose flesh nor of visceral stirrings, were present. . . . Can one fancy the state of rage and picture no ebullition in the chest, no flushing of the face, no dilation of the nostrils, no clenching of the teeth, no impulse to vigorous action, but in their stead limp muscles, calm breathing, and a placid face."

Nevertheless, I would argue that all those reactions must start in your brain *before* your body reacts to them, to *cause* that "impulse to vigorous action."

Student: But then, why should your body react to them at all?

The expressions of rage that James depicts (including that clenching of teeth and flushing of face) could have served in primordial times to help to repel or intimidate the person or creature that one is angry with; indeed, any external expression of one's mental state can affect how someone else will think. This suggests an idea about what we mean when we use our most common emotion-words; *they refer to classes of mental conditions that produce external signs that make our behaviors more predictable to the persons with whom we are dealing.* Thus, for our ancestors, those bodily signs served as useful ways to communicate such so-called "primary" emotions as Anger, Fear, Sadness, Disgust, Surprise, Curiosity, and Joy.

Student: Perhaps this could also be because our most common emotions evolved long ago when our brains were simpler. Then there were
fewer levels between our goals and our sensory-motor systems.

The body and face could also serve as a simple sort of memory: those
states of mind might soon fade away—except that those body expressions
could help to maintain those states of mind by sending signals back to the
brain. For without such "mind-body" feedback loops, the "cold and neutral" mental states that William James described might not persist for long
enough to grow into larger-scale cascades. In other words, your external
expressions of anger may serve not only to frighten your enemies, but to
also ensure that *you will stay frightened for long enough to carry out some
actions that might save your life.*

For example, your face might display an expression of horror—even
when no one else is present—when you realize that you left the door
unlocked, or forgot to turn the oven off, or that something that you
believed was false. After all, you need your body to stay alive, so, given
that it is always at hand, it makes sense for your brain to exploit it as a
dependable external memory device.

When we are young, we find it hard to suppress those external expressions, but eventually we learn to control most of them, to at least some
degree, so that our neighbors can't always see how we feel.

Student: If those physical symptoms are not essential parts of
emotions, then how can we distinguish between emotions and
other Ways to Think?

We have many names for emotional states, whereas most of our other
Ways to Think (such as those we described in Section 7-4) do not have
popular names at all—perhaps because we have not developed adequate
ways to classify them. However, here is one ancient but still useful view
of what distinguishes the mental conditions that we tend to describe as
emotional:

Aristotle b: "The emotions are all those feelings that so change
men as to affect their judgments, and that are also attended by
pain or pleasure."

In a modern version of this, some psychologists talk about *"Valence,"* which refers to the extent to which one's attitude toward some thing or situation is generally positive or negative. (See Ortony 1988.) Similarly, there is a popular view in which we think of emotion and thoughts as complementary, in much the same way that an object's color and shape can change independently; we can thus think of each object (or idea) as having various "matter-of-fact" or neutral aspects that, somehow, are also "colored" by additional characteristics that seem to make it attractive, exciting, or desirable—versus disgusting, dull, or repulsive.

More generally, our language and thoughts are filled with distinctions like "positive vs. negative" and "rational vs. emotional." Such pairs are so useful in everyday life that it's hard to imagine replacing them—any more than we should discard the idea that the sun rises and sets each day and night, although we know that this is because the earth rotates.

In particular, exaggerating the body's role in emotions can lead to serious misconceptions. Do the talents of pianists reside in their fingers? Do artists see with talented eyes? No: there is no evidence to suggest that any of those body parts think; it's the brain that sits in the driver's seat, as we see in the lives of Stephen Hawking or Christopher Reeve.


## 7-7 **Poincaré's Unconscious Processes**

> "We cannot kindle when we will
> The fire which in our heart resides,
> The spirit bloweth and is still,
> In mystery our soul abides:
> But tasks in hours of insight will'd,
> Can be through hours of gloom fulfill'd."
> —*Matthew Arnold*

Sometimes you'll work on a problem for hours or days, as when Joan worked on her progress report.

> *Joan has been thinking about her report for days but has not invented a good enough plan. Discouraged, she sets those thoughts aside . . . but then an idea "occurs" to her.*

But did Joan really set those thoughts aside, or did they continue in other parts of her mind? Hear a great mathematician recount some similar experiences:

> Henri Poincaré 1913: "Every day I seated myself at my worktable, stayed an hour or two, tried a great number of combinations and reached no results."

Most persons might get discouraged with this—but Poincaré was inclined to persist:

> "One evening, contrary to my custom, I drank black coffee and could not sleep. Ideas rose in crowds; I felt them collide until pairs interlocked, so to speak, making a stable combination. By the next morning . . . I had only to write out the results, which took but a few hours."

Then he describes another event in which his thinking seemed far less deliberate:

> "The changes of travel made me forget my mathematical work. Having reached Coutances, we entered an omnibus to go some place or other. At the moment when I put my foot on the step the idea came to me, without anything in my former thoughts seeming to have paved the way for it. . . . I went on with a conversation already commenced, but I felt a perfect certainty."

This suggests that the work was still being pursued, hidden away in "the back of his mind"—until suddenly, as though "out of the blue," a good solution "occurred" to him:

> "There was one [obstacle] however that still held out, whose fall would involve the whole structure. But all my efforts only served at first the better to show me the difficulty. . . . [Some days later,] going along the street, the solution of the difficulty that had stopped me suddenly appeared to me. . . . I had all the elements and had only to arrange them and put them together."

In the essay from which these quotations come, Poincaré concluded that when making his discoveries, he must have used activities that typically worked in four stages like these:

*Preparation:* activate resources to deal with this particular type of problem
*Incubation:* generate many potential solutions
*Revelation:* recognize a promising one
*Evaluation:* verify that it actually works

The first and last of these stages seemed to involve the kinds of high-level processes that we characterized as highly reflective ones—whereas incubation and revelation usually proceed without our being aware of them. Around the start of the twentieth century, both Sigmund Freud and Henri Poincaré were among the first to develop ideas about "unconscious" goals and processes. Poincaré suggested clearer descriptions of these (but, only for mathematical activities) and here are some of his ideas about the stages of solving difficult problems.

**Preparation:** To prepare oneself for a particular problem, one first may need to "clear one's mind" from other goals—for example, by forgetting your troubles by taking a walk, or by finding a quieter place to work. One also can try to "focus one's mind" by becoming more deliberate—as with *"It's time to sit down and start making a plan,"* or *"I must concentrate on this problem."* And because one can't solve a hard problem all at once, you'll have to break it down into smaller parts so that you can start to decide which of its features are the important ones.

Of course, this doesn't solve the problem; instead, as Poincaré said, "All my efforts only served at first the better to show me the difficulty." Nevertheless, this helps to make progress because, before you can start to solve a problem, you need to find appropriate ways to represent the situation—just as you must identify the parts of a puzzle before you can start to put them together. And until you understand the relationships among those parts, you end up wasting too much of your time at making meaningless combinations of them—as poet-critic Matthew Arnold said:

Matthew Arnold 1865: "This creative power works with elements, with materials; what if it has not those materials, those elements, ready for its use? In that case it must surely wait till they are ready."

In other words, blind trial and error won't often suffice; you need to impose constraints that generate plausible things to try.

**Incubation:** Once the "unconscious mind" is prepared, it can consider large numbers of combinations, searching for ways to assemble those fragments to satisfy the required relations. Poincaré wonders whether we do this with a very large but thoughtless search—or if it is done more cleverly.

> Poincaré 1913: "If the sterile combinations do not even present themselves to the mind of the inventor . . . does it follow that the subliminal self, having divined by a delicate intuition that [only certain] combinations would be useful, has formed only these, or has it rather formed many others which were lacking in interest and have remained unconscious?"

In other words, Poincaré asks, how selective are our unconscious thoughts? Do we explore a huge number of combinations, or work on the finer details of fewer ones? In either case, when we incubate, we will need to switch off enough of our usual Critics to make sure that the system will not reject too many hypotheses. However, we still know almost nothing about how our brains could conduct such a search, nor why some people are so much better at this.

> Aaron Sloman 1992: "The most important discoveries in science are not discoveries of new laws or theories, but the discovery of new ranges of possibilities, about which good new theories or laws can be formed."

**Revelation:** When should incubation end? Poincaré suggests that it continues until some structure is formed "whose elements are so harmoniously disposed that the mind can embrace their totality while realizing the details." But how does that subliminal process know when it has found a promising prospect?

> Poincaré 1913: "It is not purely automatic; it is capable of discernment; it has tact, delicacy; it knows how to choose, to divine.

What do I say? It knows better how to divine than the conscious self, since it succeeds where that has failed."

He conjectures that this ability to detect promising patterns seems to involve such elements as symmetry and consistency.

> Poincaré 1913: "What is it indeed that gives us the feeling of elegance in a solution, in a demonstration? It is the harmony of the diverse parts, their symmetry, their happy balance; it is all that introduces order, all that gives unity, that permits us to see clearly and to comprehend at once both the ensemble and the details."

Poincaré did not say much more about how those detectors of "elegance" might work, so we need more ideas about how we recognize those signs of success. Some of those candidates could be screened with simple matching tricks. Also, as part of the preparation phase, we select some specialized Critics that can recognize progress toward solving our problem, and keep these active throughout incubation.

**Evaluation:** We often hear advice that suggests that it's safe for us to trust our "intuitions"—ideas that we get without knowing how. But Poincaré went on to emphasize that one cannot always trust those "Revelations."

> Poincaré 1913: "I have spoken of the feeling of absolute certitude accompanying the inspiration . . . but often this feeling deceives us without being any the less vivid, and we only find it out when we seek to put on foot the demonstrations. I have especially noticed this fact in regard to ideas coming to me in the morning or evening in bed while in a self-hypnagogic state."

In other words, the unconscious mind can make foolish mistakes. Indeed, later Poincaré goes on to argue that it often fails to work out the small details—so when a "Revelation" suggests a solution, your "Evaluation" may find it defective. However, if it is only partially wrong, you may not need to start over again; by using more careful deliberation, you may be able to repair the incorrect part, without changing the rest of that partial solution.

I find Poincaré's scheme very plausible, although we surely also use other techniques. However, many thinkers have maintained that the process of creative thinking cannot be explained in *any* way, because they find it hard to believe that powerful, novel insights could result from purely mechanical processes—and hence require additional, magical talents. Moreover, some theorists question the existence of this sort of unconscious processing, and engineer Paul Plsek has summarized some such objections:

> Paul Plsek 1996: "Some experts dismiss the notion that creativity can be described as a sequence of steps in a model. For example, Vinacke 1952 is adamant that creative thinking in the arts does not follow a model, and Gestalt philosophers like Wertheimer 1945 assert that the process of creative thinking . . . does not lend itself to the segmentation implied by the steps of a model. But while such views are strongly held, they are in the minority. . . . In contrast to the prominent role that some models give to subconscious processes, Perkins 1981 argues that subconscious mental processes are behind *all* thinking and, therefore, play no extraordinary role in creative thinking."

After Poincaré, some similar models of thinking were proposed in Hadamard 1945, Koestler 1964, Miller 1960, and Newell and Simon 1972—the latter two in more computational terms. Perhaps the most extensive study of ways to generate ideas is that of Patrick Gunkel (2006). In every such model, each new proposed idea is then evaluated by activating appropriate critics. Then, if the result still has some defects, one then applies similar cycles to each apparent deficiency. In any case, it seems to me that what we call "creativity" is not simply an ability to generate completely novel conceptions; for a new idea to be useful to us, we must be able to combine it with the knowledge and skills we already possess—so it must not be too different from ideas with which we're already familiar.

## Collaboration

We usually think about thinking as a solitary activity that happens inside a single mind. However, some people excel at making new ideas, while others do better at refining them—and wonderful things can happen when the right pairs of such persons collaborate. It is said that T. S. Eliot's poetry owed much to Ezra Pound's editing, and that Sullivan's music was most inspired when working with Gilbert's librettos. We see another example of this in the Nobel Prize autobiographies of Konrad Lorenz and Niko Tinbergen:

> Niko Tinbergen: "From the start 'pupil' and 'master' influenced each other. Konrad's extraordinary vision and enthusiasm were supplemented and fertilized by my critical sense, my inclination to think his ideas through, and my irrepressible urge to check our 'hunches' by experimentation—a gift for which he had an almost childish admiration."
>    —1973 Nobel lecture

> Konrad Lorenz: "Our views coincided to an amazing degree but I quickly realized that he was my superior in regard to analytical thought as well as to the faculty of devising simple and telling experiments. . . . None of us knows who said what first, but it is highly probable that the [concept of] innate releasing mechanisms . . . was Tinbergen's contribution."
>    —1973 Nobel lecture

For many people, thinking and learning is largely a social activity—and many of the ideas in this book came from discussions with students and friends. Some such relationships are productive because they combine different sets of aptitudes. However, there are also pairs of partners who have relatively similar skills—perhaps the most important of which are effective tricks for preventing each other from getting stuck.

## Do We Normally Think "Bipolarly"?

The processes that Poincaré described involved cycles of searching and testing in which problems are solved over hours, days, or even years. However, many events of everyday thinking persist for just a few seconds or less. Perhaps these, too, begin by spawning ideas, and next selecting some promising ones, and then dwelling on their deficiencies!

Thus, suppose that a typical moment of commonsense thinking begins with a brief "micro-manic" phase that produces a few ideas; one could then quickly look for flaws in these, during a short-lived "micro-depressive" phase. If all this takes place so quickly that your reflective systems don't notice it, then each "micro-cycle" would seem to be no more than a typical moment of everyday thinking—and the entire process of thinking might seem to go in a steady, smooth, uneventful flow.[4]

The quality of each Way to Think would partly depend on how much time is spent in each phase. For example, when one is inclined to be "skeptical," one might shorten the "Incubation" phase and spend more time at "Evaluation." But if anything went badly wrong with how those durations were controlled, then (as we noted in Chapter 3-5) some of those phases might last for so long that they might appear as symptoms of a "manic-depressive" type of disorder.

## 7-8  Cognitive Contexts

No matter what you are trying to do, other temptations may attract your attention. Most such distractions can be ignored, but not when your task interrupts itself because one of its subgoals must first be achieved, or you must deal with some other emergency. Then you must put your present job on hold while switching to some other Way to Think that may require you to use some other resources and bodies of knowledge.

But after that matter has been resolved, how can you get back to your original job without having to start all over again? To do this, you'll need to reconstruct some aspects of your previous state of mind, which might include these kinds of ingredients:

> Your previous goals and priorities
> The representation you used for them
> The bodies of knowledge you had engaged
> The sets of resources that were active then
> The Selectors and Critics that were involved

This means that our model of mind needs places to store these various kinds of contextual knowledge. Otherwise, each "train of thought" would be disrupted by every interruption. In simpler brains, it might suffice to maintain only a single such memory. However, minds that look several steps ahead, or work with elaborate subgoal trees, must be able to rapidly switch among several different context-sets—because each step or subgoal may need different ways to represent its current state. So as our human minds grew more complex, we needed to evolve more machinery to enable all those processes to keep track of their different contexts.

In popular folk-psychology, we simply imagine all that stuff to be stored in our "short-term memories"—as though we could put such things into a box and take them out whenever we want. However, this image is too simple because we know that different parts of each person's brain are involved with different forms of memory—which are sometimes classified under such names as *sensory, episodic, autobiographical, semantic, declarative,* and *procedural.* Chapter 8-7 will discuss some possible forms in which those types of memory might be stored, but at present we still know very little about the structures that human brains actually use. So here we'll just ignore such details and simply imagine that all those records are stored in various parts of what we'll call the *"context box."*[5]

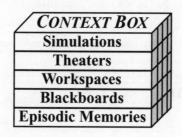

If you asked Joan what she was thinking about in the episode described in Section 7-1, she might mention the subject of tidying up. Further questions might reveal that she was maintaining several different representations of the changes she was planning to make—and, to enable her to switch among these, she must be able to store and retrieve various kinds of structures that describe

Her current collection of subgoal trees
Some records of recent external events
Some descriptions of recent mental acts
Her presently active fragments of knowledge
Simulations used to make her predictions

This means that Joan's context box for "tidying up" must keep track of various aspects of that task.

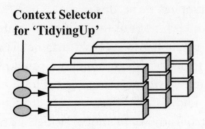

**Context Selector
for 'TidyingUp'**

Also, of course, other topics and subjects have been "on her mind" for longer spans of time, so she'll need to keep track of several such contexts, not only at different levels but also in different mental realms.

**Context Selectors for Various Jobs**

Progress Report

Tidying Up

Trip to New York

Why would we need such elaborate systems to keep track of our mental contexts? To us, it seems perfectly natural that after any brief interruption—such as to answer a question that someone has asked, or to pick up a tool that you have just dropped—we can usually get back to our previous states without needing to start all over again. It is the same when we interrupt *ourselves* to attend to a subgoal of a task, or briefly to think in some different way.

When such a diversion is small and brief, this causes little trouble because it leaves most of our active resources unchanged. However, a larger-scale change could cause more disruption and result in wasted time and confusion. So as we evolved more ways to think, we also evolved machinery for more quickly returning to previous contexts. This figure starts to combine all these ideas, to show a system in which there are similar structures for several different levels and realms.

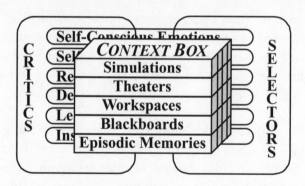

In everyday life, all these functions proceed so automatically that we have almost no sense they're happening, and we answer most questions about such matters by saying that we're just using our "short-term memories." However, any good theory of how this might work must also answer questions like these:

**How Long Do Recent Records Persist, and How Do We Make Room for New Ones?** There must be more than one answer to that because various parts of the brain must work in somewhat different ways. Some memories may be permanent, while others may rapidly fade away, unless they happen to get "refreshed." Also, some records would get erased if stored in a "place" that has a limited size—because each new item might have to replace some records that are already there. Indeed, one thing that makes modern computers so fast is that, whenever data is created or retrieved, it first is stored in what is called a "cache"—a device that has been designed to be especially quickly accessible. Then, whenever such a cache gets full, its oldest records get erased—although some of them may have been copied to larger, more permanent memory boxes.

**How Do Some Memories Become Permanent?** There is evidence that it takes hours or days for *short-term* memories to be converted to *longer-term* ones. Most previous theories about this assumed that frequent repetitions made the original record more permanent. However, it seems more likely to me that new memories are briefly maintained in resources that act like a computer's cache—and then, over time, more permanent versions are created in other regions of our brains. See Chapter 8-4.

In any case, some memories seem to last for the rest of one's life. However, this could be an illusion because they might need "refreshment" from time to time. Thus, when you recall a childhood memory, you often also have the sense of having remembered the same thing previously; this makes it hard for you to know whether you have retrieved an original record or merely a later copy of it. Worse yet, there now is ample evidence that those records can be changed while they're being refreshed.[6]

**How Do We Retrieve Old Memories?** We all know that our memories often fail—as when we try to recall some important details but find that their records have disappeared, or that at least we cannot retrieve them right now. Clearly, if no trace of that record remains, further search would be a futile quest. Nevertheless, we frequently manage to find some clues that we can use to reconstruct more of those memories. Here is a very old theory of this:

> Augustine 397: "But what happens when the memory itself loses something, as when we forget anything and try to recall it? . . . Perhaps the whole of it had not slipped out of memory; but a part was retained by which the other lost part was sought for, because the memory realized that it was not working so smoothly as usual, hence, it demanded the restoration of what was missing. For example, suppose we see or think of some man we know, and, having forgotten his name, try to recall it—but some other thing presents itself, which was not previously associated with him; then this is rejected, until something comes into the mind which better conforms with our knowledge."

In other words, once you manage to link a few of those fragments together, you may be able to reconstruct a good deal more:

> Augustine 397: ". . . by gathering together those things that the memory already contains but in an indiscriminate and confused way . . . [so that] where they formerly lay hidden, scattered, or neglected, they now come easily to present themselves to the mind which is now familiar with them."[7]

Augustine soon turned to other concerns, and concluded this discussion of memory by plaintively asking, *Who will work this out in the future?* But more than a thousand years were to pass before there was much further progress on theories about how our memories work.

# How Many Thoughts Can You Think at Once?

How many feelings can you feel at once? How many different objects or ideas can you simultaneously "pay attention" to? How many contexts can be active at once in your context box? To what extent can you be aware of how many mental activities?

The answers to such questions depend on what we mean by "aware" and "attention." We usually think of "attention" as positive, and highly regard those persons who are able to "concentrate" on some particular thing, without getting distracted by other things. However, we could also see "attention" as negative—because not all of our resources can function at once, so there is always a limit to the range of things that we can think simultaneously. Nevertheless, we can train ourselves to overcome at least some of those built-in constraints.

In any case, in our high-level thinking we can maintain only a few different "trains of thought" before we start to become confused. However, at our lower reactive levels, we carry on many different activities. Imagine that you are walking and talking among your friends while carrying a glass of wine:[8]

> Your grasping resources keep hold of the cup.
> Your balancing systems keep the liquid from spilling.
> Your visual systems recognize things in your path.
> Your locomotion systems steer you around those obstacles.

All this happens while you talk, and none of it seems to require much thought, although dozens of processes must be at work to keep that fluid from spilling out—while hundreds of other systems work to move your body around. Yet few of these processes "enter your mind" as you roam about the room—presumably because they are operating in separate realms (or separate portions of your brain) whose resources don't come into conflict with the subject you're most actively "thinking about."

It is much the same with language and speech. You rarely have even the faintest sense of what selects your normal response to the words of your friends, or which ideas you choose to express—nor of how any of your processes work to group your words into phrases so that each gets smoothly connected to the next. All this seems so simple and natural that you never wonder how your context box keeps track of what you have already said—as well as to whom you have said those things.

What limits the number of contexts that a person can quickly turn on and off? One very simple theory would be that each context box has a limited size, so there is only a certain amount of room in which to store such information. A better conjecture would be that each of our well-developed realms acquires a context box of its own. Then, some processes in each of those realms could do work on their own, without getting into conflicts until they have to compete for the same resources.

For example, it's easy to both walk and talk because these use such different sets of resources. However, it is much harder to both speak and write (or to listen and read) simultaneously, because both tasks will compete for the same language resources. I suspect such conflicts get even worse when you think about what you're thinking about, because every such act will change what is in the context box that tries to keep track of what you were thinking about.

At our higher reflective levels, our representations span many scales of time and space that can range from thinking *"I'm holding this cup"* to *"I am a mathematician,"* or *"I am a person who lives on the Earth."* To be sure, a person may have the impression of thinking all these simultaneously, but I suspect that these are constantly shifting and that our sense of thinking them all at once comes partly from the Immanence Illusion we discussed in Chapter 4-1, because the contents of our various context boxes are so rapidly accessible.

## What Controls the Persistence of Processes?

> Edmund Burke 1790: "He that wrestles with us strengthens our nerves and sharpens our skill. Our antagonist is our helper. This amicable conflict with difficulty obliges us to an intimate acquaintance with our object and compels us to consider it in all its relations. It will not suffer us to be superficial."

Whatever you happen to be doing right now, you could have chosen other alternatives, and whatever you're trying to think about, you have other concerns that compete with it. So we all have thoughts and feelings like these:

> "I've been spending so much time on this problem that I am losing my motivation; besides, it has gotten so complex that I simply cannot keep track of it; perhaps I should quit and do something else."

When none of the methods we've tried have worked, how much longer should we persist? What determines when we should quit—and lose whatever investment we've spent? We always have at least some concern with how we conserve our materials, energy, money, and friends—and each such concern would seem to suggest that we have some Critics that detect when each particular element may be getting into short supply, and then suggest ways to conserve or replenish it. Such Critics would lead us to think, *"I'm doing too many things at once,"* or *"I can't afford to buy both of these,"* or *"I don't want to lose my friendship with Charles."*

The simplest way to conserve your time is to abandon the goals that consume too much of it. But renouncing goals will often conflict with your ideals, as when they are jobs that you've promised to do; then you might want to suppress those ideals, or even regard them as handicaps. However, going against your high-level values can lead to cascades that you recognize as Tension, Guilt, Distress, or Fear—along with the Shame and Humiliation we talked about in Section 7-3. So making such decisions can thus cause you to become "emotional."

> Citizen: But certain well-disciplined persons seem able to set such emotional feelings aside, and simply do what seems "rational." Why do most other people find this so hard to do?

It seems to me that it's only a myth that there exists any single "rational" way to think. One is always comparing various goals, and deciding which ones to put aside or postpone—and one will never make much progress toward achieving any particular goal unless one can persist at it for long enough. This means that every Way to Think will need at least some ability to keep other processes from stopping it—and this could be done to some extent by controlling which Critics are active. In Chapter 3-7 we discussed some reasons why we must not keep all our Critics on all the time, and here are a few more issues involved with this:

> **What if your set of active Critics does not change?** Then you would likely keep repeating the same approach because, after each attempt to change your Way to Think, those Critics would try to switch you back again—and you might get stuck with a "one-track mind."

*What if some Critics stay on all the time?* Certain Critics must always be active to make us react to serious hazards—but if these are not selected carefully, it could lead to obsessive behaviors by making you focus too much of the time on only a few particular subjects.

*What if all your Critics get turned off?* Then all your questions would seem to be answered because you are no longer able to ask them, and all your problems would seem to be gone because nothing would seem to have any flaws.

Everything may seem wonderful during such a "mystical experience," but such revelations usually fade when enough of your Critics get turned back on.

*What if too many Critics are active at once?* Then you'd keep noticing flaws to correct and spending so much time repairing them that you would never get any important things done, and your friends may perceive you as depressed.

*What if too many Critics are turned off?* If you can ignore most alarms and concerns, that would help you to "concentrate," but it also might lead you to ignore many errors and flaws. However, the fewer Critics you activate, the fewer goals you will try to pursue, and that could make you become too mentally dull.

*What if your Critics too often get switched?* Your thinking would become chaotic if too many goals were to freely compete without any larger-scale management.

Then what should determine which Critics are active? Sometimes we need to concentrate, but we must also respond to emergencies—and all this suggests that it would be dangerous for us to have a single, centralized system that too firmly controls which of our Critics are active.

Generally, low-level selections have briefer effects, as when one of your block-building Critics insists, *"Make sure that your elbow won't topple that block."* This will alter your short-term tactics without changing your overall strategy—and even if you should make a mistake, you may be able to make a correction and continue with your original plan.

However, higher-level failures lead to larger-scale changes in strategies—for example, by arousing self-reflective thoughts that cause you to "brood" about what the future might hold for you or about your social relationships, as in, *"I don't have enough self-discipline,"* or *"My friends will lose their respect for me,"* or *"I lack the aptitude to solve this kind of problem. Perhaps I should switch to a different profession."* Such processes could lead to the large-scale cascades that people call "emotional."

## 7-9 Central Problems for Human Psychology

"No problem is so formidable that you cannot walk away from it."
—*Charles Schulz*

Much recent research in psychology has been more concerned with how large groups of people behave than about the particular ways in which each individual person thinks. In other words, at least in my view, those studies have become too statistical. I find this disappointing because, in my view of the history of psychology, far more was learned, for example, when Jean Piaget spent several years observing the ways that three children developed, or when Sigmund Freud took several years to examine the thinking of a rather small number of patients.

> Statistical psychologist: But when you study a sample so very small, then you may come to conclusions that won't generalize to larger populations. This puts you in danger of finding rules that apply only to that particular few.

I'm afraid that the opposite danger is worse, because this statistical type of research can miss good ideas about how any particular person works—thus overlooking the small but vital details. For example, when psychologists ask, *"Does passively watching violence in films make people more aggressive in real life?"* some statistical studies suggest that there is only a small correlation between these.

However, this can lead to a wrong conclusion, if one also goes on to assume that *a small correlation implies a small effect*—because a small correlation can appear when two or more different large effects happen to cancel each other out![9] The trouble is, that kind of information can simply disappear in statistical research—unless those studies look more closely to show how different individuals may use different Ways to Think about the very same kinds of situations.

It seems to me that this is why, although statistical methods were highly productive in early experiments on animals, they rarely led to good, new ideas about the levels at which only people can think. This is why I want to emphasize the importance of trying to classify the Types of Problems that people recognize, the Ways to Think that we develop, and how we learn which Ways to Think can help us to deal with each of those different Problem Types. Here are a few low-level Problem Types:

> Some obstacle is in the way.
> My goal didn't achieve its supergoal.
> I don't have access to knowledge I need.
> One of my predictions has failed.
> Two of my subgoals seem incompatible.
> I cannot get this method to work.

Many Problem Types occur at higher, more reflective levels.

> This problem is too difficult. (Divide it into smaller parts.)
> I can't. (Switch to a different representation.)
> I can't control the resources I need. (Stop to think, and reorganize.)
> This situation keeps repeating. (Switch to a different method.)
> I can't think of any worthwhile goals. (Become depressed.)
> I'm losing track of what I'm doing. (Recognize a cause for confusion.)

Similarly, whenever we switch between different Ways to Think, we also must switch among contexts like these:

> Collections of subgoals with different priorities
> Allocations of time and effort to spend
> Particular ways to represent each situation
> Ways to detect progress on each problem
> Particular ways to make predictions
> Ways to find analogies with similar problems

All this suggests that if we want to better understand the higher levels of human thought, we should ask our researchers—both in AI and in psychology—to put higher priorities on discovering ways to describe and classify the Problem Types that people face, the Ways to Think we use to deal with them, and the higher-level organizations we use to manage our mental resources. The lack of good theories about subjects like these could be why our bookshelves are filled with so much advice about ways for people to help themselves. It seems to me that this demonstrates the need for more research on questions like these, to discover more about how our everyday thinking works.

> What are the principal Problem Types that our mental Critics
>    recognize?
> What are the major Ways to Think that our mental Selectors
>    engage?
> How are our brains organized to manage all those processes?

Here is how William James once tried to depict what happens when he tries to think:

> William James 1890: "I am aware of a constant play of furtherances
> and hindrances in my thinking, of checks and releases, tendencies
> which run with desire, and tendencies which run the other way . . .
> welcoming or opposing, appropriating or disowning, striving with
> or against, saying yes or no."

Chapter 8 will talk about some of the features that give human thinking its resourcefulness, and Chapter 9 will suggest some ideas about how all those abilities might combine to form the things that we call "minds."

# 8
# RESOURCEFULNESS

## 8-1 Resourcefulness

> Descartes 1637: "Although machines can perform certain things
> as well as or perhaps better than any of us, they infallibly fall short
> in others, from which we may discover that they did not act from
> knowledge, but only from the arrangements of their parts."

We all are accustomed to using machines that are stronger and faster than
people are. But before the first computers appeared, it was hard to see how
any machine could do more than only one particular kind of task. Perhaps
this was why Descartes went on to say that no machine would ever be as
resourceful as a person can be.

> Descartes 1637: "For while reason is a universal instrument which
> can apply to every situation, a machine's parts need a particular
> arrangement for each particular action; therefore it is impossible for
> a single machine to have enough diversity to enable it to act in all
> the events of life in the same way as our reason causes us to act."

In earlier eras there also seemed to be unbridgeable differences between
the capacities of humans and other animals. Thus, in *The Descent of Man,*
Darwin observes, "Many authors have insisted that man is divided by an
insuperable barrier from all the lower animals in his mental faculties."

However, Darwin suggests that this difference may be merely a matter of degree.

> Charles Darwin 1871: "It has, I think, now been shewn that man and the higher animals, especially the primates . . . all have the same senses, intuitions, and sensations,—similar passions, affections, and emotions, even the more complex ones, such as jealousy, suspicion, emulation, gratitude, and magnanimity; . . . they possess the same faculties of imitation, attention, deliberation, choice, memory, imagination, the association of ideas, and reason, though in very different degrees."

Then he observes that "the individuals of each species may graduate in intellect from absolute imbecility to high excellence," and argues that even the highest forms of human thought could have developed from such variations—because he sees no particular point at which this would meet any intractable obstacle.

> Charles Darwin 1871: "That such evolution is at least possible, ought not to be denied, for we daily see these faculties developing in every infant; and we may trace a perfect gradation from the mind of an utter idiot . . . to the mind of a Newton."

Nevertheless, we would still like to know more details about the sequence of transitional steps that led from animal minds to human ones. In fact, there are still people who insist that such changes must have been too complex to be found by small yet useful variation. However, it appears to me that most of those skeptics are ignorant of this astonishing yet simple fact:

> It needs only a few small structural changes to vastly increase what simple computing machines can achieve. This was not known until 1936, when Alan Turing discovered how to make a "Universal" computer—that is, a single machine which, all by itself, could do all the things that all other computers can possibly do.

Specifically, Alan Turing showed how to make a machine that can inspect a description of any other machine—and then interpret that description as rules for doing just what that other machine would do.[1]

Also, then we could make such a machine to remember descriptions of several other machines and then—by switching among those different descriptions—that same machine can, step by step, do all that those other machines can do.

In other words, Turing showed how a single, "Universal" machine could use many different Ways to Think—and today, all modern computers use that very same trick of storing descriptions of other machines. (In fact, those are just what "computer programs" are.) This is why we can use the same computer to arrange our appointments, edit our texts, or help us send messages to our friends. Furthermore, once those descriptions are stored *inside* a computer, we can also write programs that can change other programs so that the machine can use those new programs to keep extending its own abilities. This showed that the limits which Descartes observed were not inherent in machines but resulted from our old-fashioned ways to build or to program them. For until our modern computers appeared, each machine that we built in the past had only one way to accomplish its task, whereas each person, when stuck, has alternatives.

Nevertheless, many thinkers still maintain that machines can never achieve such feats as composing great theories or symphonies. Instead, they prefer to attribute such feats to inexplicable "talents" or "gifts." However, those abilities will seem less mysterious once we see how our resourcefulness could result from having multiple Ways to Think. Indeed, each previous chapter of this book discussed some way in which our minds provide such alternatives:

Chapter 1. We are born with many kinds of resources.
Chapter 2. We learn from our Imprimers and friends.
Chapter 3. We also learn what we ought not to do.
Chapter 4. We can reflect upon what we are thinking about.
Chapter 5. We can predict the effects of imagined actions.
Chapter 6. We use huge stores of commonsense knowledge.
Chapter 7. We can switch among different Ways to Think.

This chapter discusses yet additional features that make human minds so versatile.

Section 8-2. We can see things from many points of view.
Section 8-3. We have ways to rapidly switch among these.

Section 8-4. We have developed special ways to learn very quickly.
Section 8-5. We learn efficient ways to retrieve relevant knowledge.
Section 8-6. We keep extending the range of our Ways to Think.
Section 8-7. We have many different ways to represent things.
Section 8-8. We develop good ways to organize these representations.

At the start of this book, we noted that it is hard to conceive of ourselves as machines, because no machine that we've seen in the past seemed to understand the meanings of things. Some philosophers argue that this must be because machines are merely material things, whereas meanings exist in the world of ideas, which lie outside the realm of physical things. However, Chapter 1 suggested that we, ourselves, have constrained our machines by defining those meanings so narrowly that we fail to express their diversity:

> If you "understand" something in only one way, then you scarcely understand it at all—because when something goes wrong, you'll have no place to go. But if you represent something in several ways, then when one method fails, you can switch to another. That way, you can turn things around in your mind to see them from different points of view—until you find one that works for you!

To show how this kind of diversity makes human thinking so versatile, we'll start by discussing the multiple methods that people use to estimate our distance from things.

## 8-2 Estimating Distances

Why has not man a microscopic eye?
For this plain reason, man is not a fly.
Say what the use, were finer optics giv'n,
T' inspect a mite, not comprehend the heav'n?
—*Alexander Pope, in* Essay on Man

When you're thirsty, you look for something to drink—and if you notice a nearby cup, you can simply reach out to pick it up—but if that cup lies farther away, then you will have to move over to it. *But how do you know which things you can reach?* A naïve person sees this as no problem at all, because "you just look at a thing and you see where it is." But when Joan detected that oncoming car in Chapter 4-2 or grasped that book in Chapter 6-1, *how did she know its distance from her?*

In primeval times we had to guess how close our predators were to us; today we only need to judge if we have enough time to cross the street—but, still, our lives depend on this. Fortunately, we each have many different ways to estimate the distance to things.

For example, you know that a typical cup has about the size of a human hand. So if a cup fills as much of the scene as does your outstretched hand, then you can reach it from where you stand. Similarly, you can judge how far you are from a typical chair because you already know its approximate size.

However, even when you don't know an object's size, you still have ways to estimate its distance from you. For example, if you can assume that two things are of similar size, then the one that looks smaller is farther away. Of course, that assumption may be wrong, if one of those objects is a small model or toy. And also, whenever two objects overlap, then the one in front must be closer to you, regardless of its apparent size.

You can also get spatial information from how the parts of a surface are lighted or shaded, and from an object's perspective and context. Again, such clues are sometimes misleading; the images of the two blocks at the right below are identical, but the context suggests that they have different sizes.

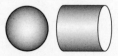               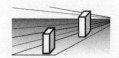

If you assume that two objects lie on the same level surface, then the one that looks higher lies farther away. Also, finer-grained textures look farther away, and so do things that look hazier.

 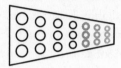

You can also judge a distance to an object by the difference between the directions from it to your two eyes or by the small differences between two images.

 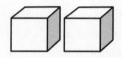

In addition, if an object is moving, then the closer it is to you, the faster it will appear to move. You can also estimate its range by how you must change the focus of the lens of your eye.

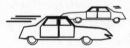 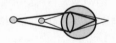

Finally, aside from all these perceptual schemes, one frequently knows where an object is without using any vision at all—because if you've seen that thing in the recent past, its location is still in your memory!

Student: Why would we need so many different methods, when surely just two or three would suffice?

In almost every waking minute, we make hundreds of judgments of distance, and yet we scarcely ever fall down the stairs or accidentally walk into doors. Yet each of our separate ways to estimate distance has many different ways to fail. Focusing works only on nearby things—and many persons can't focus at all. Binocular vision works over a longer range, but quite a few people are unable to compare the images in their two eyes. Some methods fail when the ground isn't level, and texture and haze are not often available. Knowledge applies only to objects you know, and an object might have an unusual size—yet we scarcely ever make fatal mistakes because we can use so many different techniques.

But if every method has virtues and faults, how do we know which ones to trust? The next few sections will discuss some ideas about how we manage so quickly to switch among so many Ways to Think.

Student: Why would we actually need to switch? Why can't we use all those methods at once?

There are always limits to how many things a person can do simultaneously. You can touch, hear, and see things concurrently because those processes use different parts of the brain. But few of us can draw two different things at once with both hands—presumably because these compete for resources that can do only one of those things at a time.

## 8-3 Panalogy

We have seen how useful it is to know many different ways to achieve the same goal. However, switching between alternatives could slow us down, unless we had ways to do it rapidly. This section will describe some machinery that our brains might use to do such switching almost instantly.

For example, when you read, "Charles gave Joan the book," in Chapter 6-1, you could have interpreted "book" in different realms—for example, as a physical object, a person's possession, or as a public storehouse of knowledge. Yet when you shift between those realms, that same sentence tells three different stories, as "Joan" changes from a spatial location into a recipient of a gift, and then into a person who is likely to read that book. Furthermore, you can switch between those meanings so quickly that you have almost no sense of doing this.[2] In Chapter 6-1, we introduced the term *panalogy* to describe a scheme in which corresponding features of different meanings are connected to the very same parts of just one larger structure.

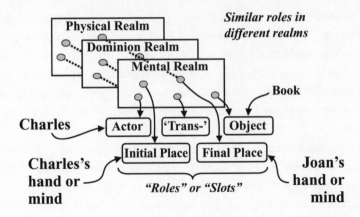

*Similar roles in different realms*

Now walk to the west wall that is now to your left, and turn yourself to face to the right; then you will be facing toward the east.

Similarly, you can think of an automobile as a vehicle, or as a complex mechanical object, or as a valuable possession—and you can regard a city as a place for people to live, as a network of social services, or as an object that requires water, food, and energy. Chapter 9 will argue that, whenever you think about your Self, you are reflecting about a panalogy of mental models of yourself.

I suspect that we use the same technique for understanding visual scenes. For example, whenever you walk into a room, you expect to see the opposite walls, but you know that you will no longer see the door through which you entered that room.

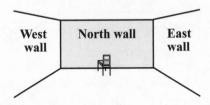

Now walk to the west wall that is now to your left, and turn yourself to face to the right; then you will be facing toward the east.

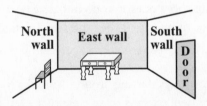

The south wall has now come into view, and the west wall is now in back of you. Yet although that wall now is out of sight, *you have no doubt that it still exists*. What keeps you from believing that the south wall just now began to exist, or that the west wall has actually vanished? This must be because you assumed all along that you are in a typical boxlike room. So of course you knew just what to expect: all four sides of that room will still exist.

Now consider that each time you move to another place, every object that you have seen may now project a different shape onto the retinas in your eyes—*and yet those objects do not seem to have changed!* For example, although the visual shape of that north wall has changed, you still see it as rectangular. What makes those meanings remain the same?[3] Similarly, you now see an image of that chair in which it appears to have turned around—but you usually don't even notice this, because your brain knows that it is *you* who has moved and not the chair. Also, you now can see the door through which you entered—yet none of this surprises you!

What if you next turn right to face the south? Then the north wall and chair will disappear, and the west wall will reenter the scene—just as anyone would expect.

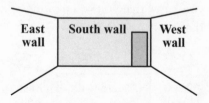

You are constantly making these kinds of predictions without any sense of how your brain deals with that flood of changing appearances: *How do you know which things still exist? Which of them have actually altered their shapes? Which of them have actually moved? How do you know that you are still in the same room?*

> Student: Perhaps those questions do not arise because we're seeing those objects continuously. If they suddenly changed, we'd notice this.

In fact, our eyes are always darting around, so our vision is far from continuous.[4] All this evidence seems to suggest that even before you entered that room, you have already, somehow, assumed a good deal of what you were likely to see.

> Minsky 1986: "The secret is that sight is intertwined with memory. When face to face with someone you newly meet, you seem to react almost instantly—but not as much to what you see as to what that sight "reminds" you of. The moment you sense the presence of a person, a world of assumptions are aroused that are usually true about people in general. At the same time, certain superficial cues remind you of particular people you've already met. Unconsciously, then, you will assume that this stranger must also resemble them, not only in appearance but in other traits as well. No amount of self- discipline can keep those superficial similarities from provoking assumptions that may then affect your judgments and decisions."

What would happen if every time you moved, you had to re-recognize every object in sight? You would have to reguess what each object is, and get evidence to support that conjecture. If so, then your vision would be so unbearably slow that you would be virtually paralyzed! But clearly this is not the case, because:

> Minsky 1974: "When we enter a room, we seem to see the entire scene almost instantly. But, really, it takes time to see—to apprehend all the details and see if they confirm our expectations and beliefs. Our first impressions often have to be revised. Still, how could so many visual cues so quickly lead to consistent views? What could explain the blinding speed of sight?"

Answer: We don't need to constantly "see" all those things because we build virtual worlds in our heads. Hear one of my favorite neurobiologists:

> William H. Calvin 1966: "The seemingly stable scene you normally 'see' is really a mental model that you construct—the eyes are actually darting all around, producing a retinal image as jerky as an amateur video, and some of what you thought you saw was instead filled in from memory."

We make those mental models so fluently that we feel no need to ask ourselves how our brains construct them and put them to use. However, here we need a theory about why, when our bodies move, the objects around us seem to remain in place. When first you see the three walls of that room, you might have represented them with a network like this:

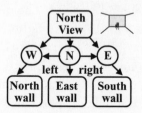

However, that representation is incomplete because, even before you entered that room, you expected it to have *four* walls—and already knew how to represent a *typical boxlike four-walled room.* Consequently, you "assumed by default" that its edges, corners, ceiling, and floor would be parts of *a larger, nonmoving framework that doesn't depend on your present point-of-view.* In other words, the "reality" that we perceive is based on mental models in which things don't usually change their shapes or disappear, despite their changing appearances. We mainly react to what we expect—and tend to represent the things that we see as though they remain the same as we move around.[5]

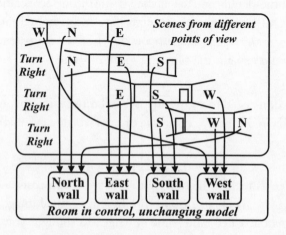

If you use the kind of larger-scale structure seen in the previous figure, then as you roam about that room, you can store each new observation in some appropriate part of that more stable framework. For example, if you represent that chair as *near* the north wall, and the door as *part* of the south wall, then these objects will have a fixed "mental place"—regardless of where you were when you noticed them—and those locations will stay the same even when those objects are out of sight. (Of course this could lead to accidents, if an object was moved without your knowing it!)

For vision, this shows how the space that surrounds us would seem to stay the same when we see it from different views—by linking features in different realms to similar roles in a larger-scale frame.

We rarely make an entirely new idea; instead, we usually modify an existing one, or combine some parts of some older ones. For, before you make records of anything new, it is likely that you have already recalled some similar object or incident—so then you can copy and modify some structure that you already possess. This is especially useful because if you were to construct an entirely new mental structure, you also would have to construct some way to retrieve it in future times, as well as to connect it to some skills for using it. However, if that older object or incident already belongs to a panalogy, and you add your new concept as an additional leaf, then it will inherit all the techniques by which your older ideas are retrieved and applied.

For example, you can think of a chair as a physical structure whose parts consist of a back, seat, and legs. In that physical view, the chair's legs support its seat, and both of these support the chair's back. You also can think of a chair as a way to make people feel comfortable. Thus, the chair's seat is designed to support one's weight, the chair's back serves to support one's back, and the chair's legs support one up to a height designed to help a person to relax.

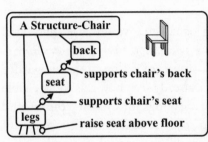
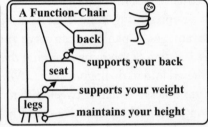

Similarly, you could also regard that very same chair as an item of personal property, or as a work of art or of carpentry—and each of those various contexts could lead you to represent chairs in different ways. Then, when your present idea of that chair makes no sense, your Critics could tell you to switch to a different mental realm—and if you have linked similar features into panalogies, that switching could work very rapidly.

> Student: How do we make those panalogies? How hard are they
> to construct and maintain? Is the talent for making them innate
> or learned? How do we learn to make use of them? Where do we
> place them in our brains?

I suspect that we do not need to "learn" all those skills, because the architecture of our brains has evolved to have structures that make it easy for us to link every fragment of knowledge we learn to ones that we already know, in similar structures for other realms and for the same things seen from different points of view. We do all this so automatically that it seems to require no reasoning; however, Section 8-5 will argue that intelligent learning requires a good deal more machinery than was imagined in most older theories about psychology.

> Student: But wouldn't such links cause you to mistake what you
> see for something else that it reminds you of? You would always
> be confusing things.

Yes, and we're constantly making those kinds of "mistakes"—but paradoxically, that often helps to keep us from being confused! For if you saw each chair as entirely new, then it would have no meaning to you. However, if each new chair reminds you of similar ones—then you will see many uses for it.

Representing knowledge by using panalogies would have substantial advantages. A panalogy can serve as a way *to use the very same structure to serve several purposes,* by changing the contexts or realms of the knowledge entered into its analogous "slots." We've already seen how this could enable us to quickly switch between different meanings for the same things, and how each such perspective could help to overcome some deficiencies of the other ones. More generally, this would be a straightforward way to represent a great many kinds of metaphors and analogies. All this has led me to suggest that our brains might embody much of our commonsense knowledge in the form of panalogies.*

> If our memories mainly consist of panalogies, then most of our thoughts will involve ambiguities. However, this is a virtue and not a fault—because much of our human resourcefulness comes from using analogies that result from this.

## 8-4 How Do People Learn So Rapidly?

Long ago, the philosopher Hume raised the question of why we can learn at all:

> David Hume 1748: "All inferences from experience suppose, as their foundation, that the future will resemble the past, and that similar powers will be conjoined with similar sensible qualities. If there be any suspicion that the course of nature may change, and that the past may be no rule for the future, all experience becomes useless, and can give rise to no inference or conclusion."

In other words, learning itself can work only in a suitably uniform universe. But still we need to ask how learning works—and we especially want to know more about how human learning works, because no other creatures come close to us in being able to learn so much. Furthermore, we do this with astonishing speed, as compared to other animals—so here we'll focus on the question of how a person can learn so much from seeing a single example.[6] Here is an episode that illustrates this:

> Jack saw a dog do a certain trick, so he tried to teach that trick to his own pet, but Jack's dog needed hundreds of lessons to learn it. Yet Jack learned that trick from seeing it only once. How did Jack so quickly learn so much—although he had only seen one instance of it?

---

*The idea of a panalogy was suggested in Section1-6 of Minsky 1974, and more details were proposed in Chapter 25 of Minsky 1986. I don't know of any experiments to see if structures like these can be found in our brains. Finding them might be difficult if different realms have representations in far-apart regions of our brains because this would require long nerve connections between analogous slots of these knowledge frames.

People sometimes need long sessions of practice, but we need to explain those occasions in which we learn so much from a single experience. However, here is a theory which suggests that Jack does, indeed, need to make many repetitions—but he does them by using an "animal trainer" inside his head, and he uses this "trainer" to train other resources inside his brain, in much the same way that he, himself, would teach his pet!

To do this, Jack could use a process like the Difference-Engine in Chapter 6-3. It would begin with his description of that trick, which is now in his short-term memory. Then Jack's "mental animal trainer" would work to produce a copy of that description in some other, more permanent place—by repeatedly altering the new copy until the trainer can see no significant difference between those short-term and long-term memories. We could do this by making a very small change in the process described in Chapter 6-3:

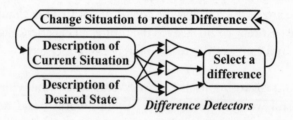

To convert this to be a copying machine, we could simply design it to *change the second description instead of the first*—until the structure in long-term memory appeared to be the same as the one in the short-term memory.[7]

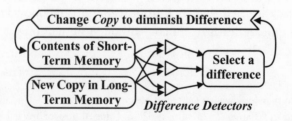

A "MENTAL ANIMAL TRAINER"

Of course, if Jack's description of that trick is composed of many smaller parts, then this cycle of changing the copy will need many repetitions.[8] So our "animal-trainer" theory about how we make new long-term memories does suggest that, in this respect, humans *are* like other animals and *do* need multiple repetitions. However, we are rarely aware of this, presumably because that process goes on in parts of the brain that our reflective thinking can't "see."

> Student: Why can't we simply remember things by making those short-term records more permanent—in the same places where they are already stored? Why should we need to copy them to other places in our brains?

This is a matter of economics: our short-term memories are limited because they use expensive resources. For example, most persons can repeat a list of five or six items, but when there are ten or more items, we reach for a writing pad. Presumably, that capacity is limited because each of our fast-access "memory boxes" consists of so much specialized machinery that each brain includes only a few of them. So we would lose a precious short-term memory box each time we made its connections more permanent!

It is probably no coincidence that modern computers evolved in a similar way: at every stage of development, fast-acting memory units were far more costly than slower ones. So the computer designers invented "caches" that use expensive, fast-acting devices only to store information that is likely to soon be needed again. Each modern computer has several such caches that work at different speeds, and the faster each is, the smaller it is—and presumably, this is also what happens inside our brains.

This would account for the well-known fact that whatever we learn is first stored temporarily—and then it may take an hour or more to convert it into a more permanent form.[9] Thus a blow to the head can cause a person to lose all memory of what happened before and including that accident. Indeed, that "transfer to long-term memory" process sometimes takes a whole day or more, and may require substantial intervals of sleep. (See Stickgold 2000.)

Here are a few other reasons why the formation of longer-term memories may have evolved to require so much time and processing.

**Retrieval:** After one makes a memory record, it would make no sense to store this away without providing some ways to retrieve it. This means that each record must also be made with links that will help to activate it when relevant (for example, by linking each new memory to some other, already existing panalogy).

**Credit Assignment:** A memory record of how one solved a problem would be unlikely to have much future use if it applied to only one situation. Section 8-5 will discuss some techniques we might use to extend the relevance of our memory records.

**The "Real-Estate" Problem for Long-Term Memories.** How could an "animal trainer" find room in the brain for the copy that it is trying to make? How could it find appropriate networks of brain cells to use without disrupting connections and records that one would not want to erase? Finding places for new memories must involve complex constraints and requirements, and this could be a reason as to why making permanent records takes so much time.

**Copying Complex Descriptions.** It is easy to imagine ways to record a simple list of symbols or properties, but I have never seen any plausible schemes for how a brain could quickly make copies of structures with more complex connections. That is why this section proposed using a sequential, Difference Engine–like scheme. (Chapter 22 of Minsky 1986 suggests that a similar scheme must be involved in verbal communication.)

## How Does Human Learning Work?

The word *learning* is useful in everyday life—but when we look closely, we see that it includes many ways that our brains can change themselves. To understand how our minds grow, we would need ideas about how people learn such different skills as how to build a tower or tie a shoe, or how to understand what a new word means, or how to guess what their friends are thinking about. If we tried to describe all the ways in which we learn, we'd find ourselves starting a very long list that includes techniques like these:

Adding new *If→Do→Then* rules
Changing low-level connections

Making new subgoals for goals
Choosing better search techniques
Changing high-level descriptions
Making new Suppressors and Censors
Making new Selectors and Critics
Linking older fragments of knowledge
Making new kinds of analogies
Making new models and virtual worlds

As children, we not only learn particular things, but we also acquire new thinking techniques. However, no infant could ever invent, by itself, enough to develop an adult intelligence. So perhaps our most important skill is how we learn, not only from having our own experiences, but also *from being told things by other people.*

## 8-5 Credit-Assignment

To the optimist, the glass is half full.
To the pessimist, the glass is half empty.
To the engineer, the glass is twice as big as it needs to be.

When first we met Carol in Chapter 2, she learned to use spoons for moving fluids. But then we asked about which aspects of her several attempts should get credit for her final success:

Should her learning include which shoes she wore, or the place in which those events occurred, or whether the weather was cloudy or clear? Which of the thoughts she was thinking then should be recorded in what she remembers? What if she smiled while using that fork, but happened to frown when using that spoon? What keeps her from learning irrelevant rules, like "To fill a cup, it helps to frown"?[10]

Some early ideas about how animals learn were based on schemes in which each reward for a success will cause a small "reinforcement" of some connections in the animal's brain—while every disappointment will cause a corresponding weakening. In simple cases, such a scheme could enable a brain to select the right features to recognize. However, in more complex situations, such methods will not work so well to find which features are relevant—and then we'll need to think more reflectively.

Some other theories about how learning works assumed that this consisted of making and storing new *If*→*Do* reaction-rules. This could be one reason why Jack's dog in Section 8-4 needed so many repetitions: perhaps, each time that dog attempted that trick, it made a small change to some *If* or some *Do*—but then, it only recorded that change in the case that it got a reward.

Simply adding new *If*→*Do* rules might suffice for learning how to do simple things—but even this may require one to make some critical kinds of decisions. Consider that any new *If*→*Do* rule is likely to fail if the *If* specifies too few details (because then that rule will be applied too recklessly)—or if the *If* includes too many details (since then it may never apply again, because no two situations are ever exactly the same). The same applies to the *Do* of that rule; therefore, each new *If* and *Do* must be just abstract enough to make it apply to a "similar" case—but not to too many dissimilar ones. Otherwise, Jack's dog might need a different new rule for every posture or place that it's in. All this means that those old "reinforcement" schemes might explain some of how certain animals learn, but those ideas won't help much to explain how humans learn more complicated things.

This brings us back to that question about how *a person can learn so quickly, without doing so many repetitions*. Earlier we suggested that we actually do many repetitions, but that these go on later inside our minds. But here we'll take another view in which we use higher-level processes to decide what to learn from each incident—when, to understand what caused your success, you need to reflect on your recent thoughts. Here are a few of the processes that making such "credit assignments" might involve.[11]

Choosing how to represent a situation will affect which future
    ones will seem similar.
Learn only the parts of your thinking that helped, and forget those
    that were irrelevant.
Connect each new fragment of knowledge so that you can access
    it when it is relevant.

The better those decisions are made, the more you will benefit from each experience. Indeed, the quality of our credit assignments could be important aspects of the suitcase of traits that people call "intelligence." For merely recording solutions to problems will help us only to solve somewhat similar problems, whereas if we can record *how we found* those solutions, that could further enable us to deal with much broader classes of situations.

For example, in playing a game like checkers or chess, if you should happen to win a game, you won't gain much by simply recording the moves that you made—because you're unlikely ever again to encounter those same situations. However, you can do better if you can learn which of your higher-level decisions helped to reach those winning positions. For as Allen Newell observed fifty years ago,

> Allen Newell 1955: "It is extremely doubtful whether there is enough information in "win, lose or draw," when referred to the whole play of the game [so, for learning to be effective], each play of the game must yield much more information. . . . If a goal is achieved, its subgoals are reinforced; if not they are inhibited. . . . Every tactic that is created provides information about the success or failure of tactic search rules; every opponent's action provides information about success or failure of likelihood inferences and so on."

Thus, when you finally achieve a goal, you should assign some credit for this to the higher-level method you used to divide that goal into subgoals. Instead of just storing solutions to problems, you can use each such experience to refine the strategies that you used.

> Student: But then you'd also want to remember the strategies that led to *those* strategies—and you've started a process that never will end!

There is no clear limit to how long one could dwell on what might have led to a certain success. Indeed, such realizations are sometimes delayed for minutes, hours, or even days (as we saw in Chapter 7-7). This suggests that some of our credit assignments involve extensive searches that go on in other parts of our minds.

For example, we sometimes have "revelations" like *"Now I see the solution to this,"* or *"I suddenly see just why that worked!"* But as we saw in Chapter 7-7, we cannot assume that those problems were solved at just those particular moments of time, since we were unaware of the unconscious work that had preceded them. If so, then such an event may merely celebrate the moment at which some Critic has said, *"This has taken so long that it's time to stop—and to adopt the tactic already considered which, at this moment, would seem the best."* [12]

We usually make our credit assignments without much reflection, but sometimes one may say to oneself after completing some difficult job, *"It was stupid of me to have wasted all that time, when I knew how to do it all along."* To remedy this, one might be able to construct a new Critic, or make some change in an existing Critic that failed to remind one to retrieve that particular fragment of knowledge.

However, such self-reflections often fail because one finds it harder to see how one found a solution than it was to solve that problem; this happens when we don't know enough about how our own mental processes work. In other words, our ability to "introspect" is limited; if it were not, we'd have no need for psychologists. So if we want to understand how people learn, we will need more research on questions about what kinds of credit assignments infants can make, how children develop better techniques, how long such processes persist, and about the extent to which we can learn to control them. In Chapter 9 we will also discuss how our feelings of pleasure might relate to how we make our credit assignments.

**Transfer of Learning to Other Realms.** Every teacher knows the frustration that comes when a child learns something to pass a test, yet never applies that skill to anything else. What makes certain children excel at "transferring" knowledge to other, different realms—whereas other children seem to need to relearn the same ideas in each domain?

It would be easy simply to say that some children are "more intelligent"—but that would not help us to explain how they use their experiences to make more helpful generalizations. This could be partly because some children are better at making and using panalogies. But also, those "smarter" children may have come to learn more efficiently because they have learned to reflect (perhaps unconsciously) about how their own learning processes work—and then found ways to improve those processes. For example, such reflections may lead to better ideas about which aspects of things they *ought* to learn.

It seems clear that the qualities of how we learn must depend, to a large extent, on how well we make our credit assignments. This means that persons who do not learn to make good credit assignments would be likely to show deficiencies in their ability to apply what they learn to new situations. This is what psychologists call *"Transfer of Learning."*[13]

This section has argued that, to gain more from each experience, it would not be wise for us to remember too many details—but only those aspects that were relevant to our goals. Furthermore, what we learn can be more profound if we assign the credit for our success not only to the final act that has led to our failure or success—or even to the strategy that has led to it—but to whatever earlier choices we made that selected our winning strategy. Our abilities to make good credit assignments could be among the most important ways in which we surpass our animal relatives.

## 8-6 Creativity and Genius

"The best way to have a good idea is to have lots of ideas."
—*Linus Pauling*

We admire our Einsteins, Shakespeares, and Beethovens—and many people insist that their accomplishments are inspired by "gifts" that cannot be explained. If so, then machines could never do such things because (at least, in that popular view) no machine could hold any such mysteries.

However, when one has the fortune to meet one of those persons whom we portray as "great," one finds no single, unusual trait that seems to account for their excellence. Instead (at least it seems to me), all that we find are unusual combinations of otherwise common ingredients.[16]

They are highly proficient in their fields. (But by itself we just call this expertise.)

They have more than usual self-confidence. (Hence better to withstand the scorn of peers.)

They often persist where others would quit. (But others may just call this stubbornness.)

They accumulate more Ways to Think. (But then they'll need better ways to switch.)

They habitually think in novel ways. (But so do others, albeit less frequently.)

They have better systems for self-control. (So they waste less time on irrelevant goals.)

They reject many popular myths and beliefs. (Especially about what
    *cannot* be achieved.)
They tend to keep thinking more of the time. (They spend less effort
    on unproductive ideas.)
They excel at explaining what they've done. (So their work is less
    likely to be neglected.)
They tend to make better credit assignments. (So they learn more
    from less experience.)

Everyone has some share of each such trait, but few develop so many of
them to such unusually great extents.

> Citizen: Each of those traits might help to explain how regular
> people solve everyday problems. But surely there must be some-
> thing unique about such great thinkers as Feynman, Freud, and
> Asimov.

Here is a statistical argument against the belief that genius comes from
singular gifts or characteristics:

> Suppose that there were, say, twenty traits that might help to
> make someone exceptional, and assume that each person has an
> even chance to excel at each particular one. Then we'd expect only
> one in each million persons to excel at all of those twenty traits.

However, even if that argument were right, it would shed no light on why
those particular persons develop so many of those particular traits. For
example, perhaps to acquire so many such qualities, *a person must first
develop some unusually good ways to learn.* In any case, there is plenty of
solid evidence that, to a significant extent, many of our mental traits are
genetically inherited. However, I suspect that yet more important are the
effects of fortunate mental accidents. For example, most children discover
various ways to arrange their toy blocks into columns and rows—and if
observers praise what they've done, those children may go on to refine
those new skills. Then a certain few of them may also go on to play at
*discovering new Ways to Think.* However, no outside observer can see those
mental events, so those particular children must have learned good ways
to, internally, praise themselves! This means that when such a child does
remarkable things, outsiders may see no clear cause for this—and will

tend to describe that child's new skills with uninformative terms like *talents, endowments, traits,* or *gifts.*

The psychologist Harold G. McCurdy suggested this particular "fortunate accident" that could bring out exceptional traits in a child—namely, to have been born in an environment that includes exceptional parents.

> Harold McCurdy 1960: "The present survey of biographical information on a sample of twenty men of genius suggests that the typical development pattern includes these important aspects: (1) a high degree of attention focused upon the child by parents and other adults, expressed in intensive educational measures and usually, abundant love; (2) isolation from other children, especially outside the family; (3) a rich efflorescence of fantasy (i.e. creativity) as a reaction to the preceding conditions. . . . [Mass education in public schools has] the effect of reducing all three of the above factors to minimum values."

It would also appear that outstanding thinkers must have developed some effective techniques that help them to organize and apply what they learn. If so, then perhaps those skills of "mental management" should get some credit for what we perceive as the products of genius. Perhaps, once we understand such things, we'll be less concerned with teaching particular skills and more with teaching children how to develop more generally powerful mental techniques.

> Citizen: But can we really hope to understand such things? It still seems to me that there is something magical about the ways in which some people imagine completely new ideas and creations.

Many phenomena seem magical until we find out what causes them. In this case, we still know so little about how our everyday thinking works that it would premature to assume that there is any essential difference between "conventional" and "creative" thought. Then why would we cling to the popular myth that our heroes have inexplicable "gifts"? Perhaps we're attracted to that idea because, if those achievers were *born* with all their wonderful tricks, we would bear no blame for our own deficiencies—nor would those artist and thinkers deserve any credit for their accomplishments.

This section has mainly aimed to explain why some people get better ideas than others do. But what if we change that question to ask, instead, what could make one person become *less* resourceful than another one? Here is one process that could tend to limit the growth of one's versatility:

> *The Investment Principle:* If you know two different ways to achieve the same goal, you'll usually start with the method that you know best. Then, over time, that method may gain so much additional strength that you'll tend to use it exclusively—even if you have been told that the other technique is the better one.

Thus, sometimes the obstacle to learning a new Way to Think is that one needs to endure the discomfort of many awkward or painful performances. So, one "secret of creativity" may be to develop the knack of enjoying that sort of unpleasantness! We'll explore this more in Chapter 9, when we talk about "adventurousness."

Speaking of "creativity," it is easy to program a machine to spout an endless stream of things that never before were conceived. However, what distinguishes the thinkers whom we call "creative" is not how many new ideas they produce—nor even how novel those concepts may be—but how effectively they can select which new ideas to further develop. This means that those artists have ways to suppress (or, better, not even to generate) products that have too much novelty.

> Aaron Sloman 1992: "The most important discoveries in science are not discoveries of new laws or theories, but the discovery of new ranges of possibilities, about which various laws or theories could be formulated. This deepens our knowledge of the 'form' of the world, as opposed to its 'contents' or its 'constraints'—the laws."

## 8-7 Memories and Representations

> William James 1890: "There is no property absolutely essential to one thing. The same property, which figures as the essence of a thing on one occasion, becomes a very inessential feature upon another."

Everyone can imagine things; we hear words and phrases inside our minds. We envision conditions that don't yet exist—and then exploit those images to predict the effects of possible actions. Much of our human resourcefulness comes from our being able to manipulate mental representations of objects, events, and conceptualizations.

But what do we mean by a *representation*? I will use that term to talk about any structure inside one's brain that one can use to answer some questions. Of course, those answers will be useful only when your representation behaves enough like the subject that you are asking about.

We sometimes use actual physical objects to represent things, as when we use a picture or map to help us find paths between parts of a city. However, to answer a question about a past event, we must use what we call our "memories." But what do we mean by a "memory"? Each memory must be some record or trace that you made at the time of some prior event—and, of course, you cannot record an event itself; at best, you can only make some records about some of the objects, ideas, and relationships that were involved in that incident—*as well as how that event affected your mental state.* For example, when you hear a statement like *"Charles gave Joan the book,"* you might represent that incident with a scriptlike sequence of *If→Do→Then* rules:

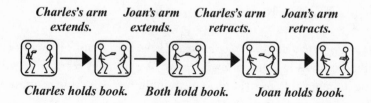

**Charles's arm**   **Joan's arm**   **Charles's arm**   **Joan's arm**
**extends.**     **extends.**     **retracts.**     **retracts.**

**Charles holds book.**    **Both hold book.**    **Joan holds book.**

However, you also may want to represent your knowledge about whether that book was a gift or a loan, or whether Charles's goal was to ingratiate Joan, or how the actors in that scene were dressed, or the meanings of some of the words that they said. So, we typically make several different representations for any particular incident. For example, those records might include:

A verbal description of that incident
A visual simulus of the scene
Some models of the persons involved

Simulations of how those persons felt
Analogies with similar incidents
Predictions about what might happen next

Why would your brain represent the same event in so many different ways? If each realm of thought that you engaged left an additional record or trace, this would enable you, later, to use multiple ways to think about that same incident—for example, by using verbal reasoning, or by manipulating mental diagrams, or by envisioning the actors' gestures and facial expressions.

Today, we still know little about how our brains make those memory traces or how they later retrieve and "replay" them. For although today we know a lot about the ways in which separate brain cells behave, we know far less about how those cells are organized into larger structures that represent our memories of past events. Nor do our own reflections tell us much about the details of those processes; usually, the most we can say is simply that we "remember" some things that have happened to us. So the following section will suggest a few structures that our brains might use to represent the knowledge stored in our memories. Then we'll go on to speculate about how such structures might be arranged in our brains.

## Multiple Ways to Represent Knowledge

*What distinguishes us from other animals?* One important difference is that no other animals ask questions like that! We humans seem to be almost unique in being able to treat ideas as though they were things or, in other words, to "conceptualize."

However, to invent new concepts and put them to use, we must represent those new ideas in structural forms that we store in the networks inside our brains—because no small fragment of knowledge can have any meaning unless it is part of some larger structure that has connections to other parts of our network of knowledge. However, it does not much matter how those links are embodied; the same computer can be made with wires and switches, or even with pulleys, blocks, and strings; all that matters is how each part changes state in response to the changes in the other parts to which it is connected.

In other words, knowledge is not composed of "ideas" that exist as separate entities that float around in some mental world; they also need to be interconnected. Of course, it often is useful to think of thoughts

and ideas as being "abstract"—and to represent them with symbols in diagrams, or with the sentences of written texts. Nevertheless, for a thought
or concept to have any effect—such as to cause your hand to move a
block, or to make your vocal tract produce a sound, or *to cause you to
think of your following thought*—there must be some physical structures
that interconnect some representations inside your brain.

This section reviews some modern ideas that researchers have used
for representing knowledge inside computers—as well as some that have
not yet been tested. There is not enough room here for many details, and
*The Society of Mind* says much more about the subjects of the rest of this
section.[14]

## Describing Events as Stories or Scripts

Perhaps our most familiar way to represent an incident is to recount it
as a story or script that depicts a sequence of events in time—that is, in
the form of a story or a narrative. We've seen such scripts for the sentence
*"Charles gave Joan the book"* and for Carol's plan about how to build
an arch.

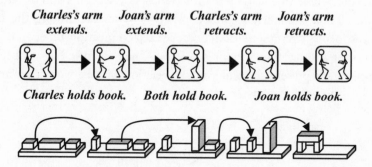

Of course, not all processes are so linear. Most computer programs are
mainly composed of sequential actions like these—but at some points the
flow is interrupt by some branching *If*s, at which the script can go in several directions, depending on specified current conditions. Nevertheless,
once the process is done, one can simplify and summarize it by listing only
the path that was actually taken—as in, *"I was trying to build an arch with
my toy blocks, and I discovered that one has to build the supports before put-
ting on the top"*—which omits any mention of all the diversions involved
in learning this.

## Describing Structures with Semantic Networks

Of course, each of the items named in a story or script may refer to other, more complex structures. For example, to understand what terms like *Joan* or *book* mean, the reader must already possess some structures or models that represent these. Whenever we need to describe more details, such as the relations between an object's parts, it may be better to use the kinds of structures we saw in Chapters 4-6 and 5-8 to represent a person or a physical book.[15]

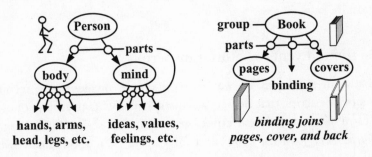

Each of these so-called *"Semantic Networks"* is a collection of symbols that are linked by labeled connection links. They are among the most versatile forms of representation, because each connection-link could itself refer to yet another type of representation. The Semantic Network in the figure below represents several kinds of relationships between the various parts of a three-block Arch.

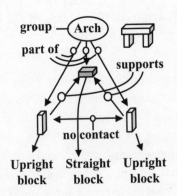

Each of the links named *part of, group, supports,* and *do not touch* refers, in turn, to some other structure, resource, or process that one can use to understand more about what this Semantic Network represents. For example, the links that bear the label "supports" could be used to predict that the top block would fall if we removed a block that supports it.

## Using Trans-Frames to Represent Actions

To represent the effects of an action, it is convenient to use pairs of Semantic Networks to represent what was changed. In Chapter 5-8 we saw how to imagine replacing the top of an arch, by changing a single name or relationship at a high level of representation—instead of altering thousands of points to change a visual picturelike image.

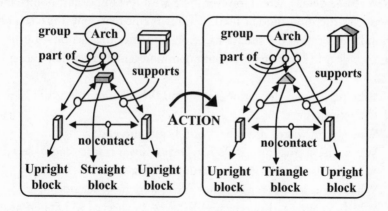

We'll use the term *"Trans-Frame"* to name a pair of representations of the conditions before and after some action was done. Then we can represent the effect of a sequence of actions by linking together a chain of Trans-Frames to form a story or narrative. We've already seen how to represent *"Charles gave Joan the book"* with five such "movie frames":

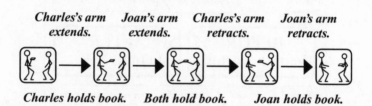

Here is another version of this consisting of only three such representations, but in these each frame shows some further details.

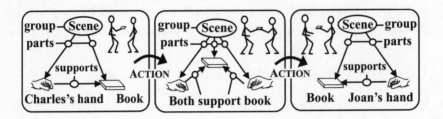

## Using Frames to Embody Commonsense Knowledge

I described Frames and Trans-Frames in Minsky 1974 and 1986, so I won't repeat all the details here. However, here are a few important points about how such structures could be used. A Trans-Frame represents the effect of an action by describing the situations before and after that action—but it can also include other information about commonsense matters like these:

> Who performed the action, and why? What other things did the
>     action affect?
> Where (and when) did that action begin and end?
> Was it intentional or not? What purposes was it intended to serve?
> What kinds of methods or instruments were used?
> What obstacles were overcome? What were its other side effects?
> Which resources did it engage? What was expected to happen next?

For example, a Trans-Frame for Joan's trip from Boston to New York could have additional "slots" like those depicted here:

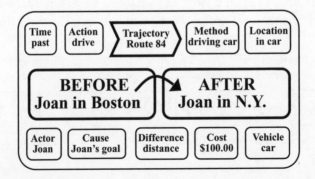

This Trans-Frame includes two Semantic Networks that describe the situations *Before* and *After* that trip was taken—but it also contains many other slots that contain information about when, how, and why Joan took that trip. But also, those slots may already come equipped to contain the common answers to the most common questions. In other words, the slots of our Frames may contain "by default" a great deal of what we call commonsense knowledge.

For example, when someone says "apple," you seem to know instantly that a typical apple grows on a tree, is round and red, is about the size of a human hand, and has a certain texture, flavor, and taste—yet almost no time seems to elapse between hearing that word and then becoming aware of such things. Chapters 6 and 7 asked questions about what could enable your brain so quickly to retrieve so much of the commonsense knowledge it needs. Our theory is that this is partly because every slot of each of your frames has already been filled with the most common or typical information. Then you use this to make a good guess whenever you don't have additional information.

For example, you might assume "by default" that an apple is red—but if you know that a certain apple is green, then you will replace "red" with "green" in its color slot. In other words, a typical frame describes a stereotype whose "default assumptions" are usually right—but which you can easily change whenever you meet some exceptions to them.[16]

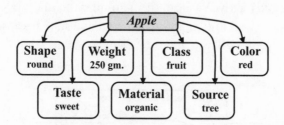

All adults know millions of items like these, and regard them as everyday, commonsense knowledge, but every child takes years to learn all the nuances of how their Trans-Frames behave under different conditions and in different realms. For example, everyone knows that, if you *move* a thing in the physical realm, then this will change the location it's in—but if you *tell* some information to your friend, then that same knowledge will be in two places at once. Similarly, if you heard that Charles was grasping a book, you would not stop to ask why he was doing that, because you would assume by default that he has the most usual goal for anyone who would grasp *anything*—that is, to keep it from falling to the floor.

This idea of "default assumptions" could help to explain how you can so quickly access your commonsense knowledge: as soon as you activate a frame, many questions that you might otherwise ask will already be answered before you can ask them.[17]

## Learning by Building "Knowledge-Lines"

Suppose that you have just had a good idea, which helped you to solve a certain hard problem called P. What should you learn from this experience? One thing you could do is to construct a new rule: *If* the problem you face is like problem P, *Then* try the solution that once worked on P. Such a rule will help you to solve problems that closely resemble P—but will be less helpful with less similar problems. However, if you could make a recording of the Way to Think that you used to find that solution, this would be more likely to help in a wider range of situations.

Of course, it would be impractical to make a copy of the entire state of a human mind; however, you might get most of the effect you want if you could, later, reactivate enough of the resources that were active at the time you discover that way to solve problem P. You could do this by constructing a new Selector that is connected so that it activates just those resources that were recently active. We call this kind of structure a *"K-line."* Such a K-line can act as a sort of "snapshot" of a mental state because, when you later activate it, this will put you into a similar state.

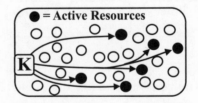

Here is an analogy that illustrates how K-lines work:

> Kenneth Haase 1968: "You want to repair a bicycle. Before you start, smear your hands with red paint. Then every tool you need

to use will end up with red marks on it. When you're done, just remember that 'red' means 'good for fixing bicycles.' If you use different colors for different jobs, some tools will end up marked with several colors. [ . . . ] Later, when you have some job to do, just activate the set of tools with the right color for that kind of job, and the resources that you've used for such jobs then become available." [See Chapter 8 of Minsky 1986.]

This way, for each kind of problem or job, your K-lines can fill your mind with ideas that might be relevant—by getting you into a mental state that resembles one that helped in the past to do a similar task.

Student: I see how this new K-line could be used as a Selector for a new Way to Think. But how would you build a new Critic to recognize when to activate it?

If we want to use that new K-line for problems similar to P, then such a Critic should recognize some combination of features of P.

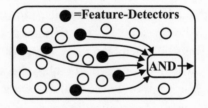

However, if such a Critic would act only when *all* the features of P were present, then it could fail to recognize situations that were slightly different from P. So each new Critic should only detect features that actually helped.

Student: I see what you mean. Suppose that when you were fixing that bicycle, at some point you tried to use a tool that happened to make the problem worse. It wouldn't be good to paint that tool red because, then, later you would waste more time again.

This suggests that when we make new Selectors and Critics—or more generally, whenever we learn—we should try to make sure that *what* we learn will mainly consist of knowledge that's likely to actually help. The following section on credit assignment will discuss some processes that might help to ensure that what we learn will be relevant in future times.

> Student: Would those K-lines help you to do anything new, if each of them merely revisits a Way to Think that you already knew how to use?

That would rarely be a problem because, when you activate a K-line, that won't completely replace your present Way to Think—because that K-line will turn some of your resources off and turn some set of others on, but many of your current resources will still remain engaged. So now, two different sets of resources will be active in your mind at once: the ones used for your recent thoughts and the ones aroused by that memory. If those resources were all compatible, then both sets might be able to work together to solve the problem that you are facing. Then you could combine what remains of both of those sets, store them away as a new K-line set—and the result would be that you would have created a Selector for a new Way to Think.

What would happen if too many of your current resources were incompatible with those that a K-line tried to activate? One strategy might be to give priority to the K-line's resources—but that policy could have bad side effects: we do not want our memories to re-create old states of mind so firmly that they overwhelm our present thoughts, because then we might lose track of our present goals or wipe out all the work that we've recently done. Another policy would be to give the presently active agents priority over the remembered ones, and yet another policy would suppress both.

My answer is that no single policy will always work; therefore, resourceful people find ways to decide (using higher-level strategies) which policy might be best to apply in various kinds of situations. In any case, whichever policy is used, the resulting state of mind will almost surely be a bit different from any state that your mind has ever been in. Thus, every new situation is likely to lead to a somewhat novel Way to Think—and if you make a "snapshot" of *that,* you will have a K-line that differs from all of your previous ones.[18]

We also should note that our mental representations almost never "start from scratch" because, whenever we make a new one, we usually do this by linking some older ones. For example, when you understood that

*Charles gave Joan the book,* your representation of this event would almost surely refer to previous representations of *Charles, Joan,* and *book* that you had constructed at earlier times. So after hearing that sentence, your state of mind will include many resources that those other concepts use.

Consequently, if you tried to make a single K-line that could re-create that mental state, that K-line might need to be connected to hundreds of thousands of other resources. However, you could accomplish much of the same effect by making a K-line that simply connects to just those three older representations of *Charles, Joan,* and *book.* Then, when you activate this new K-line at some later date, this may be sufficient to give you the sense of reexperiencing the mental event that you constructed it to represent.

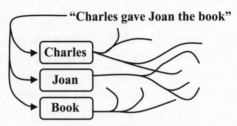

*K-line attached to three other K-lines.*

## Connectionist and Statistical Representations

Let's contrast two different ways to represent the commonsense idea of an apple—which is an edible fruit with red or yellow or green skin with a crisp whitish flesh and a sweet to tart taste, and comes from a native Eurasian tree widely cultivated in many varieties.

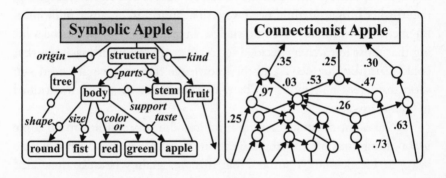

The diagram on the left shows a Semantic Network that describes various features and relationships between various aspects or parts of an apple. The diagram on the right shows an example of what is called a "*Connectionist Network*," which also displays some aspects of an apple, but does not have any simple way to distinguish between different relationships; it only shows numbers that represent how closely those features are "associated." It would take too long here to explain how such networks are used, and the reader can see more details in Minsky 1988 and Minsky 1991. Connectionist systems have had many practical applications, because they can be made to learn to recognize many important types of patterns—without any need for a person to program them.

However, those number-based networks also have limitations that keep them from doing reflective thinking. One can sometimes interpret those numerical values as correlations or likelihoods, but, because they carry no other clues about what those links might signify, it can be extremely hard for other resources to use that information. The trouble is that a Connectionist Network must reduce every relationship to a single numerical value or "strength," so there remains almost no trace of the evidence that led to it. For example, if you see only the number 12, you cannot tell if that number represents 5 plus 7, or 9 plus 3, or 27 minus 15—or whether it counts the people in a room or the legs of the chairs they are sitting on. In short, numerical representations become obstacles to using higher-level Ways to Think. In contrast, Semantic Networks can explicitly represent different kinds of relationships (because of the labels attached to each link).

I mention all this because, although I played a part in inventing Connectionist Networks, I see their popularity, in recent years, as having retarded the search for higher-level ideas about human psychological machinery. In my view of that history, research on commonsense thinking kept advancing until about 1980, but then it was clearly recognized that further progress would need ways to acquire and organize millions of fragments of commonsense knowledge. That prospect seemed so daunting that most researchers decided to try, instead, to invent machines that could learn, by themselves, all the knowledge that they would need—in short, to invent new kinds of "baby-machines" like those we mentioned in Chapter 6-2.

Quite a few of these learning machines did indeed learn to do some useful things, but none of them went on to develop higher-level reflective Ways to Think—and I suspect that this was mainly because they tried to represent knowledge in numerical terms, which made it very hard for them to produce expressive explanations.

Nevertheless, I do not mean to suggest that such networks are not important—for as we'll see in Section 8-8, it seems safe to assume that many of the *low-level* processes in our brains must use some forms of Connectionist Networks.

## Micronemes for Contextual Knowledge

We always face ambiguities. The significance of each thing that you see depends on the rest of your mental context. This also applies to events in your mind, because what they mean depends on which mental resources are active then.[19] In other words, no symbol or object has meaning by itself, because your interpretation of it will depend on the mental context you're in. For example, when you hear or read the word *block,* you might possibly think that it means an obstacle to progress, a certain kind of rectangular object, a wooden board to chop things on, or a stand on which things in an auction are shown. Then which interpretation will you select?

Such choices will depend, of course, on how your current mental context will dispose you to make selections from such sets of alternatives as these:

Conceptual or material
Well established or speculative
Robust, fragile, or repairable
Public or private
Urban, rural, forest, or farm
Irregular or symmetrical
Animal, mineral, or vegetable
Common, rare, or irreplaceable
Indoors or outdoors
Residence, office, theater, or car
Color, texture, hardness, or strength
Cooperative or competitive, etc.

Many contextual features like these have common names, but many others have no such words, just as we have no expressions for most flavors and aromas, gestures and intonations, attitudes and dispositions. I have proposed to use the term *"micronemes"* for the myriad of nameless clues that color and shade our thoughts about things, and the figure below suggests some machinery through which such contextual features could affect our mental processes.[20] Imagine that the brain contains a bundle of thousands of wirelike fibers that pass through a great many other structures inside that brain—so that the state of each of those micronemes can influence many other processes.

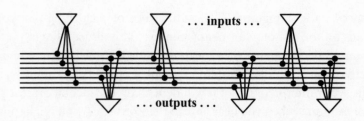

On the input side, we shall assume that many of your mental resources—such as K-lines, Frame-slots, or *If→Do→Then* rules—can alter the states of some micronemes. Then the present state of your micronemes could represent much of your current mental context—and as the states of those fibers are changed, your far-reaching bundles of micronemes will broadcast that information to many other mental resources, so that this will change some of your attitudes, outlooks, and states of mind.

## 8-8  A Hierarchy of Representations

The sections above have briefly described several kinds of structures that we could use to represent various types of knowledge. However, each of those representation types has its own virtues and deficiencies—so each of them may need other connections through which to exploit some other types of representations. This suggests that our brains need some larger-scale organization for interconnecting our multiple ways to represent knowledge. Perhaps the simplest such arrangement would be a hierarchical one.

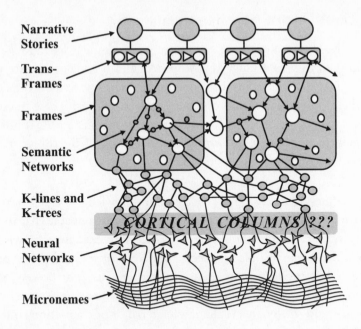

Narrative
Stories

Trans-
Frames

Frames

Semantic
Networks

K-lines and
K-trees

*CORTICAL COLUMNS ???*

Neural
Networks

Micronemes

This diagram shows one attempt to suggest how a brain might organize its multiple ways to represent knowledge. However, we should not expect to find that actual brains are arranged in such an orderly way. Indeed, we should not be surprised if anatomists find that different regions of the brain evolved somewhat different organizations to support mental functions in different realms—such as for maintaining our bodily functions, manipulating physical objects, developing social relationships, and for reflective and linguistic processes. Also, even if this diagram turned out to be a good description of how those functions relate to one another, some structures that appear to be neighbors in this picture could actually be much farther apart. Indeed, much of the mass of a human brain consists of bundles of nerves that interconnect regions at distant locations.[21]

It also seems unlikely that our representations are arranged quite so hierarchically. In biology, new structures usually originate as duplicate copies of older ones, and this often results in orderly layers. However, because brain cells are so peculiarly able to make connections to distant places, they can more easily evolve less hierarchical organizations.

## How Do We Learn New Representations?

From where do we obtain our ways to represent knowledge, and why do we find it so easy to arrange that knowledge into panalogies? Are these abilities installed genetically into our infant memory systems, or do we learn them individually from our personal experiences? These questions suggest a more basic one: how do we manage to learn at all? As Immanuel Kant pointed out long ago, *learning to learn* is one of the things that we cannot learn entirely from experience!

> Immanuel Kant 1787: "That all our knowledge begins with experience there can be no doubt. For how is it possible that the faculty of cognition should be awakened into exercise otherwise than by means of objects which affect our senses, and partly of themselves produce representations, partly rouse our powers of understanding into activity, to compare, to connect, or to separate these—and so to convert the raw material of our sensations into a knowledge of objects? [ . . . ] But, though all our knowledge begins with experience, it by no means follows that it all arises out of experience. For, on the contrary, it is quite possible that our empirical knowledge is a combination of that which we receive through impressions, and [additional knowledge] altogether independent of experience . . . which the faculty of cognition supplies from itself, sensory impressions giving merely the occasion."

So, although sensations give us "occasions" to learn, this cannot be what makes us "able" to learn, because we first must have the additional knowledge that our brains would need, as Kant has said, to "produce representations" and then "to connect" them.[22] Such additional knowledge would also include inborn ways to recognize correlations and other relationships among sensations. I suspect that, in the case of physical objects, our brains are already innately endowed with machinery to help us "to compare, to connect, or to separate" objects so that we can represent them as existing in space.

All this leads me to suspect that we must be born with primitive forms of structures like K-lines, Frames, and Semantic Networks—so that no infant needs to wholly invent the kinds of representations that we depicted above. However, I doubt that we're born with those structures complete, so it still requires some effort and time for us to refine those primitive representations into their more adult forms. I hope that there will soon be more research on how those developmental processes work.

Could any person ever invent a totally new kind of representation? Such an event must be quite rare because no type of representation would be useful without some effective skills for working with it—and a new set of such skills would take time to grow. Also, no fragment of knowledge could be of much use unless it is represented in a familiar way. For reasons like these, it makes sense to conjecture that most of our adult representations come either from refining our primitive ones, or by acquiring them from our culture. However, once a person has learned to use several different representations, then that person might be more able to invent new ones. This could be what distinguishes the work of those exceptional writers, artists, inventors, and scientists who repeatedly discover new and useful ways to represent things.

How should a brain proceed to select which representation to use? As we have emphasized several times, each particular kind of description has virtues and deficiencies. Therefore, it makes more sense to ask, "Which methods might work for the problem I'm facing—and which representations are likely to work well with those methods?"

Most computer programs still, today, can do only one particular kind of task, using only a single kind of representation—whereas our human brains accumulate multiple ways to describe each of the Types of Problems we face. However, this means that we also need to learn ways to decide which technique to use in each situation—and it also means that we need to learn how to switch to another alternative whenever the method we're using fails.

## Which Representations to Use for Which Purposes?

When programmers set out to develop a program, they usually start by selecting a way to represent the knowledge their program will need. But each representation works well only in certain realms, and none works well in every domain. Yet we frequently hear discussions like this about what is the best way to represent knowledge:

Mathematician: It is always best to express things with logic.
Connectionist: No, logic is far too inflexible to represent commonsense knowledge. Instead, you ought to use Connectionist Networks.

Linguist: No, because Connectionist Nets are even more rigid. They represent things in numerical ways that are hard to convert to useful abstractions. Instead, why not simply use everyday language—with its unrivaled expressiveness.

Conceptualist: No, language is much too ambiguous. You should use Semantic Networks instead—in which ideas get connected by definite concepts!

Statistician: Those linkages are too definite, and don't express the uncertainties we face, so you need to use probabilities.

Mathematician: All such informal schemes are so unconstrained that they can be self-contradictory. Only logic can ensure us against those circular inconsistencies.

This shows that it makes no sense to seek a single best way to represent knowledge—because each particular form of expression also brings its own particular limitations. For example, *logic-based systems* are very precise, but they make it hard to do reasoning with analogies. Similarly, *statistical systems* are useful for making predictions, but do not serve well to represent the reasons why those predictions are sometimes correct. It was recognized even in ancient times that we must represent things in multiple ways:

"[One person might describe a house] as 'a shelter against destruction by wind, rain, and heat,' while another might describe it as 'stones, bricks, and timbers'; but there is a third possible description which would say that it was that form in that material with that purpose or end. Which, then, among these is entitled to be regarded as the genuine physicist? The one who confines himself to the material description, or the one who restricts himself to the functional description? Is it not rather the one who combines both in a single formula?"
     —*Aristotle, in* On the Soul

However, sometimes it may be better *not* to combine those multiple representations.

Richard Feynman 1965: "Psychologically we must keep all the theories in our heads, and every theoretical physicist who is any good knows six or seven different theoretical representations for exactly the same physics. He knows that they are all equivalent, and that nobody is ever going to be able to decide which one is right at that level, but he keeps them in his head, hoping that they will give him different ideas for guessing."

The key word here is *guessing,* because each such theory has virtues and faults, and no single representation will be best for every predicament that we might face. So much of our human resourcefulness comes from having multiple ways to describe the same situations—so that each one of those different perspectives may help us to get around the deficiencies of the other ones. How could a person know when and how to choose any particular representation? There are several suggestions about this in my essay on Causal Diversity in Minsky 1992.

# 9
# THE SELF

Each one of us contains a set
   Of persons each will be:
Oh, how I wish my own next self
   Would take the place of me!
          —*Theodore Melnechuk*

What makes each human being unique? No other species of animal has such diverse individuals; each person exhibits a different set of appearances and abilities. Some of those traits are inherited, and some come from each person's experiences—but in every case, we each end up with different characteristics. We sometimes use *"Self"* for the features and traits that distinguish each person from everyone else.

Daniel Dennett 1991: "The strangest and most wonderful constructions in the whole animal world are the amazing, intricate constructions made by the primate *Homo sapiens*. Each normal individual of this species makes a self. Out of its brain it spins a web of words and deeds, and, like the other creatures, it doesn't have to know what it's doing; it just does it. This web protects it, just like the snail's shell. . . . As such, it plays a singularly important role in the ongoing cognitive economy of that living body, because, of all the things in the environment an active body must make mental models of, none is more crucial than the model the agent has of itself."

However, we also use Self in a sense that suggests that we are controlled by powerful beings inside ourselves, who want and feel and think for us, and make our important decisions for us. We call these our Selves or Identities—and see them as staying the same over time, regardless of what may happen to us. Sometimes we even envision that Self as a minuscule person inside the mind; this is sometimes called a *"homunculus."* (A similar premise was prevalent before the dawn of modern genetics: it claimed that every sperm already contained a perfectly formed little personage.)

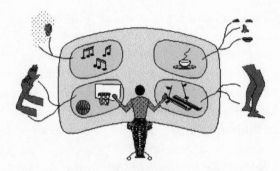

Daniel Dennett 1978: "A homunculus (from Latin, 'little man') is a miniature adult held to inhabit the brain . . . who perceives all the inputs to the sense organs and initiates all the commands to the muscles. Any theory that posits such an internal agent risks an infinite regress . . . since we can ask whether there is a little man in the little man's head, responsible for his perception and action, and so on."[1]

What attracts us to the queer idea that we can think or feel only with the help of those Selves inside our minds? Chapters 1 and 4 suggested that this concept helps to keep us from wasting time on difficult questions about our minds. For example, if you wonder how your vision works, the Single-Self view gives the answer, *"Your Self simply peers out though your eyes."* If you ask about how your memory works, you get the reply, *"Your Self knows how to recollect whatever might be relevant."* And if you wonder what guides you through your life, it tells you that your Self supplies you with all your wishes, hopes, and goals—and then solves all of your problems for you. Thus, the Single-Self view diverts you from asking about how your mental processes work. Instead, it leads you to ask questions like these:

*Is an infant born with anything like what an adult would call a "Self"?* Some would insist on answering with, "Yes, infants are persons just like us—except that they don't yet know so much." But others would take an opposite view: "An infant begins with almost no intellect, and developing one takes a sizable time."

*Does your Self have a special location in space?* Most "Western" thinkers might answer yes—and tend to locate it inside their heads, somewhere not far behind their eyes. However, I've heard that some other cultures situate Selves between the belly and chest.

*Which of your goals and beliefs are your "genuine" ones?* The Single-Self view suggests that some of your intentions and values are "authentic" and "sincere"—whereas the models of mind discussed in this book leave more room for conflicting views.

*Does your Self stay the same throughout your life?* We each have a sense of remaining the same, regardless of what may happen to us. Does this mean that some part of us is more permanent than our bodies and our memories?

*Does your Self survive the death of your brain?* Different answers to that may please or distress us, but they would not help us to understand ourselves.

Each such question uses words like *self, we,* and *us* in a somewhat different sense—and this chapter will argue that this is because, whenever we try to understand ourselves, we may need to use multiple views of ourselves.

Whenever you think about your "Self," you are switching among a network of models, each of which may help to answer questions about different aspects of what you are.

Here, as we said in Chapter 4-3, we're using the word *model* to mean a mental representation that can help us to answer some questions about some other, more complex thing or idea. For example, some of our models are based on simplistic ideas, like *"All our actions are based on the will to survive,"* or *"We always like pleasure more than pain,"* while some other Self-models are far more complex. We develop these multiple theories because each of them helps to represent certain aspects of ourselves, but is likely to give some wrong answers about other questions about ourselves.

Citizen: Why should a person want more than one model? Would it not be better to combine them into a single, more comprehensive one?

In the past, there were many attempts to make "unified" theories of psychology. However, this chapter will suggest some reasons why none of those theories worked well by itself, and why we may need to keep switching among different views of ourselves.

Jerry Fodor 1998: "If there is a community of computers living in my head, there had also better be somebody who is in charge; and, by God, it had better be Me."

"I have been reading my old poems, and they were written by somebody else. Yet I am that selfsame person; or, if I am not, who is? If no one is, when did he die—when he finished this poem, or that one, or the next day, or the end of that month?"
— *Cosma Rohilla Shalizi*

# 9-1 How Do We Represent Ourselves?

"O wad some Pow'r the giftie gie us
To see oursels as ithers see us!"
—*Robert Burns*

How do people construct their Self-models? We'll start by asking simpler questions about how we describe our acquaintances. Thus, when Charles tries to think about his friend Joan, he might begin by describing some of her characteristics. These could include his ideas about

The appearance of Joan's body and face
The range and qualities of her abilities
Her motives, goals, aversions, and tastes
The ways in which she is disposed to behave
Her various roles in the social world

However, when Charles thinks about Joan in different realms, his descriptions of her may not all agree. For example, his view of Joan as a person at work is that she is helpful and competent, but tends to undervalue herself; however, in social settings he sees her as selfish and overrating herself. What could lead Charles to make such different models? Perhaps his first representation of Joan served well to predict her social behaviors, but that model did not well describe her business self. Then, when he changed that description to apply to that business realm, it made new mistakes in the contexts where it had formerly worked. Eventually, he found that he had to make separate models of Joan to describe her behaviors in various roles.

> Physicist: Perhaps Charles should have tried harder to construct one single, unified model of Joan.

This would not be feasible, because each of a person's mental realms may need different kinds of representations. Indeed, whenever a subject becomes important to us, we tend to build multiple models for it—and this ever-increasing diversity must surely be a principal source of our human resourcefulness.

To more clearly see the need for multiple models, we'll turn to a simpler situation: suppose that you find that your car won't start. Then, to diagnose what might be wrong, you will need to switch among several different views of your car:

> If the key is stuck, or the brake won't release, you must think in terms of mechanical parts.

> If the starter won't turn, or if there is no spark, you must think in terms of electrical circuits.

> If you've run out of gas, or the air intake's blocked, you need a model of how your car burns fuel.

It is the same in every domain; to answer different types of questions, we often need different kinds of representations. For example, if you wish to study psychology, your teachers will make you take courses in at least a dozen subjects, such as neuropsychology, neuroanatomy, personality, perception, physiology, pharmacology, social psychology, cognitive psychology, mental health, child development, learning theories, language and speech, and so on. Each of those subjects uses different models to describe different aspects of the human mind.

Similarly, to learn physics, you would need to study subjects called classical mechanics; thermodynamics; vector, matrix and tensor calculus; electromagnetic waves and fields; quantum mechanics; physical optics; solid state physics; fluid mechanics; theory of groups, and relativity. Each of those subjects has its own ways to describe the events that occur in the physical world.

> Student: I thought that physicists seek to find a single model or "grand unified theory" to explain all phenomena in terms of some very small number of general laws.

Those "unified theories of physics" may be grand, indeed—but whenever we deal with complex subjects like physics or psychology, we find ourselves forced to split those domains into "specialties" that use different kinds of representations to answer different kinds of questions. Indeed, a major part of education is involved with learning when and how to switch among different representations.

Returning to Charles's ideas about Joan, these will also include some models of Joan's own views about herself. For example, Charles might suspect that Joan is displeased with her own appearance (because she is constantly trying to change it), and he also makes models of how Joan might think about herself in realms like these.

> Joan's ideas about her own ideals
> Her ideas about her abilities
> Her beliefs about her own ambitions
> Her views about how she behaves
> How she envisions her social roles

Joan would probably disagree with some of Charles's views about her, but this may not make him change his opinion, because he knows that the models that people make of their friends are frequently better than the models that people make of themselves.

> "Others often better express myself."
> —*Kevin Solvay*[2]

## We Each Make Multiple Models of Ourselves

> Greg Egan 1998: "But even as these ordinary thoughts and per-
> ceptions flowed unimpeded, a new kind of question seemed to
> spin through the black space behind them all. Who is thinking
> this? Who is seeing these stars, and citizens? Who is wondering
> about these thoughts, and these sights? And the reply came back,
> not just in words, but in the answering hum of the one symbol
> among the thousands that reached out to claim all the rest: Not to
> mirror every thought, but to bind them. To hold them together,
> like skin. Who is thinking this? I am."

We've discussed a few models that Charles might use when he thinks about
his friend, Joan. But what kinds of models might people use when they try
to think about themselves? Perhaps our most common self-model begins
(see Chapter 4-5) by representing a person as having two parts—namely,
a "body" and a "mind."

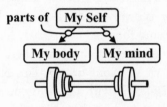

A TWO-PART SELF-MODEL

That *"body-mind"* division soon grows into a structure that describes
more of one's physical features and parts. Similarly, that part called "mind"
will divide into a host of parts that try to depict one's various mental abilities.

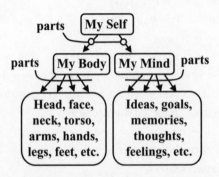

A MORE COMPLEX SELF-MODEL

Each of the models that one makes of oneself will serve well only in some situations, so one ends up with different self-portraits in which one has different abilities, values, and social roles. So when we think about ourselves, we'll usually need to keep switching among those multiple representations of ourselves.

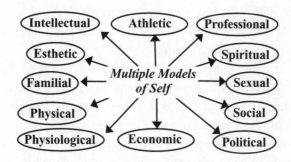

If you tried to represent all those perspectives at once, your model would soon become too complex to use; in each of those realms we portray ourselves with somewhat different autobiographies, each based on using different aims, ideals, and interpretations of the same ideas and events. Nevertheless, as Daniel Dennett suggests, we are rarely inclined to recognize this, so each of us constructs a myth of having—or being—a Single Self.

Daniel Dennett 1992b: "We are all virtuoso novelists, who find ourselves engaged in all sorts of behaviour, and we always try to put the best 'faces' on if we can. We try to make all of our material cohere into a single good story. And that story is our autobiography. The chief fictional character at the centre of that autobiography is one's self."

## Multiple Subpersonalities

"For there is not a single human being . . . who is so conveniently simple that his being can be explained as the sum of two or three principal elements. . . . Harry consists of a hundred or a thousand selves [but] it appears to be an inborn and imperative need of all men to regard the self as a unit. . . . Even the best of us shares the delusion."

—*Herman Hesse, in* Steppenwolf

When Joan is with a group of her friends, she regards herself to be a sociable person. But when surrounded by strangers, she sees herself as anxious, reclusive, and insecure. For as we said in Chapter 4-8, each person makes many different Self-models to use in each of many contexts and realms. Thus Joan's mind abounds with varied Self-models—Joans past, Joans present, and future Joans; some represent remnants of previous Joans, while others describe what she hopes to become; there are sexual Joans and social Joans, athletic and mathematical Joans, musical and political Joans, and various kinds of professional Joans.

Whenever these "subpersonalities" are actively playing their different roles, each of them may have some control over different sets of goals and skills—so that each has a somewhat different Way to Think. However, they will all need common access to many of the person's resources and bodies of commonsense knowledge. This means that those subpersonalities will frequently need to compete for control of some higher-level processes.

For example, suppose that Joan is working at her professional job, but suddenly some social part of her mind reminds her of a time when she was trapped in an awkward relationship. She tries to shake off those memories, only to find herself thinking in childish ways about how her parents would view her behavior—or she might find that she is regarding herself as a partner in a business, or as a person who likes to do research, a member of a family, a person involved in a love affair, or as a person who has a pain in her knee.

In the course of such trains of everyday thinking, we frequently switch between different Self-models, whose various outlooks may not be consistent, because we use them for different purposes. So when one needs to make a decision, the result will partly depend upon which of one's subpersonalities is active then. Joan's *Business Self* might be inclined to choose the option that seems more profitable; her *Ethical Self* might select an option that better conforms with her ideals; her *Social Self* might want to select the one that would most please her friends. For example, when we identify ourselves as members of a social group, then we can share its triumphs and failures with pride and shame—whereas, when one is involved with a business, one may feel obliged to try to suppress such sentiments. Thus, as we said in Chapter 1, each major change in emotional state may display a different subpersonality:

> When a person you know has fallen in love, it's almost as though
> someone new has emerged—a person who thinks in other ways,

with altered goals and purposes. It's almost as though a switch had been thrown, and a different program has started to run.

Whenever we switch among subpersonalities, we are likely to change our Ways to Think—but because the context remains the same, we will still maintain some of the same priorities, goals, and inhibitions, as well as some contents of short-term memories, and of our currently active mental Critics.

However, some such changes may be larger, and you often hear sensational stories about persons who switch between totally different personalities. But while such extremes are exceedingly rare, everyone undergoes changes of mood in which one exhibits somewhat different sets of intentions, behaviors, and traits. Then, whether those shifts are persistent or brief, the subpersonality that is now in control may activate a set of views and goals for you, which, for the moment, you may believe to be the views and goals of the "genuine" You.

## The Sense of Personal Identity

Augustine: "Of what nature am I? A life various, manifold, and vast. Behold in the numberless halls and caves, in the countless fields and dens and caverns of my memory, full without measure of numberless kinds of things—present there either through images as all bodies are; or present in the things themselves as are our thoughts; or by some notion or observation as our emotions are, which the memory retains though the mind feels them no longer . . ."

It sometimes makes sense to think of your Self as a permanent, unchanging entity. But to what extent are you the same as you were ten minutes ago? Or are you like the proverbial knife that has had both its handle and blade replaced? You are certainly not like the text of a bound, printed book whose "contents" don't change from one day to the next. Nevertheless, enough of your knowledge remains the same—and different enough from anyone else's—that one can argue that our Identities are mainly what's in our memories.

"A man is often willing to say that this is the same person who did something in the past, not on the basis of knowing that it

is the same body but on a quite different basis—that the person recounts the past situation with great accuracy, exhibits similar personal reactions, and displays the same skills."
    —Encyclopaedia Britannica

However, that sense of identity can fade when we change our ways to interpret our older memories.

William James 1890: "When the continuity is no longer felt,] the sense of personal identity goes too. We hear from our parents various anecdotes about our infant years, but we do not appropriate them as we do our own memories. Those breaches of decorum awaken no blush, those bright sayings no self-complacency. That child is a foreign creature with which our present self is no more identified in feeling than it is with some stranger's living child today. Why? Partly because great time-gaps break up all these early years—we cannot ascend to them by continuous memories; and partly because no representation of how the child felt comes up with the stories. . . . It is the same with certain of our dimly recollected experiences. We hardly know whether to appropriate them or to disown them as fancies, or things read or heard and not lived through. . . . The feelings that accompanied them are so lacking in the recall, that no judgment of identity can be decisively cast."

A century later, another description of what we might mean when we talk of our Selves:

Daniel Dennett 1991: "Our fundamental tactic of self-protection, self-control, and self-definition is not building dams or spinning webs, but telling stories—and more particularly concocting and controlling the story we tell others—and ourselves—about who we are. . . . And finally, we, (unlike *professional* human storytellers) do not consciously and deliberately figure out what narratives to tell and how to tell them; like spider webs, our tales are *spun by us;* our human consciousness, and our narrative selfhood, is their *product,* not their *source.* . . . These strings or streams of narrative issue forth as if from a single source—not just in the obvious physical sense of flowing from just one mouth, or one pencil or pen, but in a more subtle sense: their effect on any audience or readers is to encourage them to (try to) posit a unified agent whose words they are, about whom they are: in short, to posit what I call a 'center of narrative gravity.'"

In other words, Dennett portrays our conceptions of ourselves as like collections of drafts of self-portraits or stories that are constantly being edited by diverse assortments of processes. But then, what could you mean when you speak of yourself as remaining the same? Of course, that depends on how you're describing yourself—so, instead of asking about your Identity, perhaps you should ask, *Which of my models of myself best serves my present purposes?*" In any case, we should ask ourselves what compels us to think of ourselves as Selves—and here is a simplistic theory of this: whatever happens, we're prone to ask ourselves who or what was responsible—because our representations force us to fill the "caused-by" slots that we mentioned in Chapter 8-7. This leads us to find explanations that frequently help us to predict and control not only what happens in the world, but also what happens in our minds. Thus we often find ourselves wondering what caused us to act in a certain way or led us to make a particular choice.

However, when you fail to find a plausible cause, that slot-filling hunger may lead you to imagine a cause that doesn't exist—such as the "I" in *"I just got a good idea."* For if your frame-default machinery compels you to find some single cause for everything that you ever do—then that entity needs a name. You call it "me." I call it "you."

## 9-2 Personality Traits

Alfred Korzybski 1933: "Whatever you say something is, it is not."

If you asked Joan to describe herself, she might say something like this:

Joan: "I think of myself as disciplined, honest, and idealistic. But because I am awkward at being sociable, I try to compensate by trying to be attentive and friendly, and when that fails, by being attractive."

Similarly, if you were to ask Charles to describe his friend Joan, he might declare that she is helpful, tidy, and competent, but somewhat lacking in self-confidence. Such descriptions are filled with everyday words that name what we call "character traits" or "characteristics"—such as *disciplined, honest, attentive,* and *friendly.* But what could make it possible for someone to describe a person at all? Why should minds so complex as ours show any clear-cut characteristics? Why, for example, should anyone tend to be usually neat or usually sloppy—rather than tidy about some things but not about others? W*hy should personal traits exist at all?* Here are some possible causes for the appearance of such uniformities:

**Inborn Characteristics.** One reason why people exhibit traits is that each person is born with different sets of genes that lead to particular ways to behave.

**Learned Characteristics.** Each person also comes to learn individual goals and priorities that influence when various resources get engaged—as when to become angry or afraid—so that some individuals may tend more than others to become belligerent or diffident.

**Investment Principle.** Once we learn an effective way to do some job, we'll resist learning other ways to do it—because new methods are usually harder to use until we become proficient at them. So as our older procedures gain strength, it gets harder for new ones to compete with them.

**Archetypes and Self-Ideals.** Every culture comes with myths that describe beings endowed with well-defined traits. Few of us can prevent ourselves from becoming attached to those heroes and villains—and this makes us try to change ourselves, to make those imagined traits become real.

**Self-Control.** It is hard to achieve any difficult goal—or to carry out any long-range plan—unless you can make yourself persist at it. The following section will suggest that, to keep ourselves from constantly changing our goals and other priorities, our cultures teach us to train ourselves to become more "self-predictable," by constraining the ways in which we behave.

In any case, although our trait-based descriptions are frequently wrong and always incomplete, they help to make things seem simpler and more understandable. Thus, it is easy to say that a person is honest and tidy—as opposed to being deceitful and sloppy—no matter that no person always tells the truth, or keeps everything perfectly neat. It saves a great deal of effort and time to see people or things as stereotypes.

However, the concept of traits can be treacherous because, even when we suspect that those attributions are wrong, they may still continue to influence us. Here is a common example of this: suppose that some stranger you've never met were to take your hand, look into your eyes, and then report this impression of you:

> "Some of your aspirations tend to be unrealistic. At times you are extroverted, affable, sociable, while at other times you are introverted, wary, and reserved. You have found it unwise to be too candid in revealing yourself to others. You are an independent thinker and do not accept others' opinions without good evidence. You prefer a certain amount of change and variety, and become dissatisfied when hemmed in by restrictions and limitations.
>
> "At times you have serious doubts as to whether you made the right decision or did the right thing. Disciplined and controlled on the outside, you tend to be anxious and insecure inside. Your sexual adjustment has presented some problems for you. You have a great deal of unused capacity, which you have not turned to your advantage. You have a tendency to be critical of yourself, but have a strong need for other people to like and to admire you."[3]

Many people are amazed that a stranger could see so deeply inside of them—yet every one of those statements applies, to some extent, to just about everyone! Just look at the adjectives in that horoscope: *affable, anxious, controlled, disciplined, extroverted, frank, independent, insecure, introverted, proud, reserved, self-critical, self-revealing, sociable, unrealistic, wary.* Everyone has concerns with regard to each of those characteristics, so few of us can help but feel that each such prediction applies to us.

Thus, millions of people have been entranced by the prophecies of so-called psychics, fortune-tellers, and astrologers—even when their forecasts turn out no better than random chance would predict. (See Carlson 1985.) One reason could be that we trust those "seers" more than we trust ourselves, because they appear to be "reliable authorities." Another possible cause could be that we tend to believe that we already resemble the persons we wish to be like—and fortune-tellers excel at guessing what their clients would like to hear. However, those predictions may often ring true simply because we each maintain so many Self-models that almost any statement about ourselves will agree with at least some of those models.

## Self-Control

It is hard to achieve any difficult goals unless, at least to some extent, you can make yourself persist at it. You would never complete any long-range plans if, whatever you tried, you kept "changing your mind." However, you cannot simply "decide" to persist, because many kinds of ideas and events may later affect your priorities. Consequently, we each must develop ways to impose less breakable constraints on ourselves. In other words, we need to make ourselves predictable.

You see examples of this in how you and others construct your social relationships. Whenever you expect help from a friend, you assume that, at least to some extent, that person's behavior will be predictable. Similarly, to carry out a plan of your own, you must be able to "depend on yourself"—and so again, to that extent, you must make yourself predictable. Our cultures help us to acquire such skills by teaching us to respect such traits as commitment and consistency. For if you come to admire such traits, you may make it your goal to train yourself to behave in those ways.

> Citizen: Might not such restrictions cause you to pay the price of losing your spontaneity and creativity?

> Artist: Creativity does not result from lack of constraints, but comes from discovering appropriate ones. Our best new ideas are the ones that lie just beyond the borders we wish to extend. An expression like "skdugbewlrkj" may be totally new, but would have no value unless it connects with other things that you already know.

In any case, it is always hard to make yourself do things that do not interest you—because, unless you have enough self-control, the "rest of your mind" will find more attractive alternatives. Chapter 4-7 showed how we sometimes control ourselves by offering threats or bribes to ourselves in the form of self-incentives, like, *"I'll be ashamed of myself if I give in to this,"* or *"I'll be proud if I can accomplish this."* To do this, you need some knowledge about which of those methods will work on yourself, but generally, it seems to me, the tricks that we use for self-control are very much like the ones that we use to influence our acquaintances.

Also, we often control ourselves by exploiting things in the physical world. To stave off sleep, you can pinch yourself, or take a deep breath—

or ingest the right amount of some stimulant. Or you can move to a more exciting place, or indulge in strenuous exercise. All those activities can keep you awake by exploiting your environment. Another trick that you can use is to try to change your emotional state by assuming various postures or facial expressions: these seem peculiarly effective because *they are likely to influence you as much as they do your audience.*

But why must you use such devious tricks to select and control your Ways to Think—instead of just choosing to do what you want to do? As we said at the end of Chapter 3, *directness would be too dangerous.* You would probably die if one part of your mind could take over control of all the rest, and our species would soon become extinct if we could ignore the demands of hunger, pain, or sex. Accordingly, our systems evolved so that in emergencies, our instincts could dislodge our fantasies.

Furthermore, every culture develops ways to help its members constrain themselves. For example, every game that our children play helps train them to invent and get into new mental states that will help them obey the rules of that game. In effect, each such game is a virtual world that we use for teaching ourselves to behave in certain specified ways.

Self-control is no simple skill, and many of us spend much of our lives seeking ways to make our minds "behave." This suggests yet another meaning for Self; we sometimes use it as a suitcase-word for all the methods we use to control ourselves.

## Dumbbell Ideas and Dispositions

> There are two rules for success in life.
> First, never tell anyone all that you know.
> —*Anonymous*

Why do we find it so easy to say that a person is reclusive and shy, as opposed to being sociable—or that someone tends to be placid and calm, instead of impulsive and excitable? More generally, why do we find it so easy to make such two-part distinctions for other aspects of our personalities—as when we group our tempers, emotions, moods, and traits into pairs that we regard as opposites?

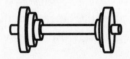

Solitary vs. Sociable
Tranquil vs. Agitated
Forthright vs. Devious
Audacious vs. Cowardly
Dominant vs. Submissive
Careless vs. Meticulous
Cheerful vs. Cranky
Joyous vs. Sorrowful

We see similar "dumbbell" thinking at work when people try to describe things in terms of opposing pairs of forces, spirits, or principles. Of course, all those distinctions are flawed; Sorrow is not the mere absence of Joy, nor is Agitation the absence of Tranquility. Nevertheless, we're all prone to divide many aspects of our minds into pairs with seemingly opposite qualities. An example of this is the popular myth that each person has two basic ways to think—that are embodied in opposite sides of the brain. In earlier times, those two halves of the brain were thought to be almost identical. But in the mid-twentieth century, when surgeons could cut the connections between those halves, some significant differences were observed, and this revived many views of the mind as a place for conflicts between pairs of antagonists like these:

Left vs. Right
Thought vs. Feeling
Rational vs. Intuitive
Logical vs. Analogical
Intellectual vs. Emotional
Conscious vs. Unconscious
Quantitative vs. Qualitative
Deliberate vs. Spontaneous
Literal vs. Metaphorical
Reductionist vs. Holistic
Scientific vs. Artistic
Serial vs. Parallel

But how could so many such distinctions be embodied in the same two halves of the very same brain? The answer is that this is largely a myth, because each of those mental activities involves the use of machinery located in both of those halves of the brain. However, there also is some truth to that myth; our brains begin as highly symmetrical but then, one side develops more machinery for language-based activities, while the other side develops more visual and spatial abilities. However, I suspect that these differences might partly result from some process in which the so-called "dominant" side develops more reflective thinking, whereas the other side remains more reactive and less deliberative. As evidence for this, Battro 2000 appears to have shown that a single half brain can do both.

Accordingly, I am inclined to conjecture that these differences might result from a process in which one side of the brain comes to develop substantially better "management skills." Of course, this could happen on both sides at once—but many conflicts would soon arise if one had to obey two masters at once. However, as soon as one side begins to excel at suppressing impulses that come from the other side, that first one could then become "dominant," while the other one might slow down in developing abilities to produce and pursue higher-level plans and goals. The result would be that the nondominant side would appear to be more childish and less mature because of having fewer administrative skills. It might need only a small genetic bias to determine which side of the brain eventually wins the prize of having more influence at the top.

Here are some other possible reasons why people like two-part distinctions so much:

**Many things seem to come in opposing pairs.** Generally, it is difficult for us to distinguish what something "is" without contrasting it with what it is not, and this makes us tend to see things in terms of their possible opposites. For example, it often makes sense to classify physical objects as *large* or *small,* or as *heavy* or *light,* or as *cold* or *hot.*

However, a young child might tell you that the opposite of *water* is *milk,* or the opposite of a *spoon* is a *fork*—but later, that very same child may also insist that the opposite of *fork* is *knife.* Thus, opposites depend on the contexts they're in, and so may overrule consistency.

**Intensities and Magnitudes.** Although it is hard to describe what feelings *are,* it seems easy to say how *intense* they are. This makes it seem quite natural to apply such adjectives as *slightly, largely,* or *extremely* to almost every emotion word—such as *sorry, pleasant, happy,* or *sad.*

We often justify a choice by declaring that we like a certain option *more* or *less* than another one—as though those options were like points on a line. However, that kind of one-dimensional comparison can lead us into supposing that both options are almost the same—except for having "plus" or "minus" signs! However, as we said, Sorrow is not the mere absence of Joy. Thus, representing feelings in terms of intensities can simplify how we make our decisions—by encouraging us to overlook other kinds of distinctions in cases where we ought to use more thoughtful ways to deal with things![4]

**Structural vs. Functional Descriptions.** Many of our distinctions are based on ways to make connections between the structures of things and the ways in which we can use those things. Accordingly, it is often convenient to classify the parts of an object as playing "principal" vs. "supporting" roles—just as we did for "a chair" in Chapter 8-3, where we identify the seat and back as its functional parts, and its legs as merely serving to hold them up.[5]

Certainly, two-part distinctions can be useful when we need to choose between alternatives, but when that fails, we may have to resort to more complex distinctions. For example, when Carol is trying to build that arch, it will sometimes suffice for her to first describe each block as being *short* or *tall*, or *narrow* or *wide*, or *thin* or *thick;* then she may need only to decide which of those distinctions is relevant. However, on other occasions, Carol may need to find a block that satisfies some more elaborate combination of constraints that relate its height, width, and depth; then she can no longer describe that block in terms of only a single dimension.

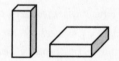

**Inborn Brain Machinery.** Another reason why we tend to think in terms of pairs could be that our brains are innately equipped with special ways to detect differences between pairs of mental representations. Thus in Chapter 6-4 we mentioned that when you touch something very hot or cold, the sensation is intense at first, but then will rapidly fade away—because our external senses mainly react to how things change in time. (This also applies to our visual sensors, but we're normally unaware of this because our eyes are almost always in motion.) If this same sensitivity to change also applies to sensors *inside* a brain, this would make it easy to compare a pair of descriptions, simply by alternately presenting them. However, this "temporal blinking" scheme would work less well for describing the relationships of more than two things—and that could be one reason why people are so much less proficient at making three-way comparisons.[6]

When is it appropriate to distinguish between only two alternatives? We often speak as though it is enough to classify a new thing or event in "yes or no" terms like these:

Was this a failure or a success?
Should we see it as usual or exceptional?
Should we forget it or remember it?
Is it a cause for pleasure or for distress?

Such two-part distinctions can be useful when we have only two options to choose among. However, selecting what to remember or do will usually depend on making more complex decisions like these:

How should we describe this event?
What links should we connect it with?
Which other things is it similar to?
What other uses could we make of it?
Which of our friends should we tell about it?

More generally, it usually makes little sense to commit ourselves, for all future times, about which objects to *like or dislike*—or about which persons, places, goals, or beliefs we should *seek or avoid,* or *accept or reject*—because all such decisions should also depend on the contexts that we find ourselves in. Accordingly, it seems to me that there is something wrong with most dumbbell distinctions: those divisions appear to be so simple and clear that they seem to be all that you need—and that satisfaction tempts you to stop. Yet most of the novel ideas in this book came from finding that two parts are rarely enough—and eventually my rule became: *when thinking about psychology, one should never start with less than three different parts or hypotheses!*

Why are we so often satisfied with dividing things into only two kinds? Perhaps this is at least partly because a typical child's environment contains so few significant "triplets" of things. A two-year-old has only *two* feet, and is taught by a *pair* of parents to learn to put on a *pair* of shoes—and soon, that typical two-year-old will learn to understand and to use the word *two*. But it frequently takes another full year for a child to learn to use the word *three*—perhaps because our environments contain so few instances of "three-nesses." We all excel at contrasting pairs of things and making lists of their differences, but our cultures and languages do not provide us with terms for talking about relationships among triplets of things. Why don't we have words for *trichotomies* or *trifferences*?

## 9-3  Why Do We Like the Idea of a Self?

> Brian: *You are all individuals!*
> Mob: *We are all individuals!*
> Lone voice: *I'm not.*
>     —*Monty Python's* The Life of Brian

Most of the time we think of ourselves as having definite identities.

> Introspectionist: I do not feel like a scattered cloud of separate parts and processes. Instead, I sense that there's some sort of Presence in me—an Identity, Spirit, or Feeling of Being—that governs and guides all the rest of me.

Other times we find ourselves feeling less unified or centralized.

> Citizen: One part of me wants this, while another part of me wants that. I need to get more control of myself.

One philosopher claimed never to feel any sense of unity.

> Josiah Royce 1908: "I can never find out what my will is by merely
> brooding over my natural desires, or by following my momentary
> caprices. For by nature I am a sort of meeting place of countless
> streams of ancestral tendency. . . . I am a collection of impulses.
> There is no one desire that is always present to me."

In any case, even when we feel that we're in control, we recognize conflicts
among our goals and compulsions that we can't overcome. Then we may
argue inside our minds, trying to find a compromise—but even when we
feel unified, others may see us as disorganized.

We solve easy problems in routine ways, scarcely thinking about how
we accomplish these—but when our usual methods don't work, we start
to "reflect" on what went wrong and find ourselves to be switching around
in a network of "models," each of which purports to represent some facet
or aspect of ourselves, so that we end representing ourselves with a loosely
connected collection of images, models, and anecdotes.

Still, if this is how one represents one's Self, there is nothing special
about this—*because that's how we represent everything else.* Thus, when you
think about a telephone, you keep switching among different views of its
appearance, its physical structure, and the feelings you have when you use
it, and so forth, as though exploring the facets of a *panalogy.* So when you
think about your Self, you are using the same techniques with which you
think about everyday things; parts of your mind keep switching among
multiple models and processes. But if so, then what impels us to believe
that we must be anything more than Josiah Royce's meetings of streams?
What leads us to the strange idea that our thoughts cannot just proceed by
themselves, but need yet something else to control themselves?

> Jerry Fodor 1998: "If there is a community of computers living in
> my head, there had also better be somebody who is in charge; and,
> by God, it had better be Me."

> Citizen: If no central Self exists, why does it feel as though I have
> one? When I think my thoughts and imagine things, must not
> there be someone who's doing those things?

Obviously, there's a problem with this: if we had those Single Selves to
*want and feel and think for us,* then we would not have any need for

minds—and if our minds could do those things by themselves, then of what possible use would those Selves be to us? *Aha!* Perhaps that is precisely the point: I suspect that we use words like *"Me"* and *"I"* to *keep us from thinking about what we are!* For they all give the very same answer, *"Myself,"* to every such question that we might ask. Here are some other ways in which that Single-Self concept is useful to us in everyday life.

**A Localized Body.** You cannot walk through solid walls, or stay aloft without support. Where any part of your body goes, the rest of you must also go—and the Single-Self model implies the idea of being in only one place at a time.

**A Private Mind.** It is pleasant to think of your Self as being like a strong, closed box, so that no one else can share your thoughts to learn the secrets you want to keep—for only you hold the keys to those locks.

**Explaining our Minds.** Perhaps it seems to make sense to say things like *"I perceive the things that I see,"* because we know so very little about how our perceptions actually work. This way, that Single-Self view can help to keep us from wasting time on questions we don't know answers to.

**Moral Responsibility.** Each culture needs behavioral codes. For example, because our resources are limited, we all agree to censure Greed. Because we each depend on others, we have agreed to chastise Treachery. And to justify our laws and decrees, we assume that our Selves are "responsible" for every willful, intentional deed.

**Centralized Economy.** We'd never accomplish anything if we kept asking questions like "Have I considered every alternative?" We prevent this with Critics that interrupt us with, "That's enough thinking; I've made my decision!"

**Causal Attribution.** When we represent any thing or event, we like to attribute some Cause to it. So when we don't know what led to some thought, we assume that the Self was the cause of it. This way, we may sometimes use the word *"Self"* the way we say *It* in *"It started to rain,"* because we don't know a more plausible cause.

**Attention and Focus.** We often think of our mental events as occurring in a single "stream of consciousness"—as though they all were emerging from some single, central kind of source, which can attend to only one thing at a time.

**Social Relations.** Other people expect us to think of them as Single Selves, so unless we adopt a similar view, it will be hard to communicate with them.

These all are good reasons why the Single-Self view is convenient to use in our everyday lives. But if you want to understand how your thinking works, no simple model could portray enough details of how our minds work. Nor would it help for you to have some way to observe your entire mind simultaneously, because then you would be overwhelmed by seeing too many unwanted details. So eventually, you will need to switch among simplified models of yourself.

Why must those models be simplifications? Each model must help us to focus on only those aspects that matter in some particular context; that's what makes a map more useful to us than seeing the entire landscape that it depicts. The same applies to what we store in our minds. Consider how messy our minds would become if we filled them up with descriptions of things whose details had too little significance. So instead, we spend large parts of our lives at trying to tidy up our minds—selecting the portions we want to keep, suppressing others we'd like to forget, and refining the ones we're dissatisfied with.[7]

## 9-4 What Is Pleasure, and Why Do We Like It?

Aristotle b: "We may lay it down that Pleasure is a movement by which the soul as a whole is consciously brought into its normal state of being; and that Pain is the opposite. If this is what pleasure is, it is clear that the pleasant is what tends to produce this condition, while that which tends to destroy it, or to cause the soul to be brought into the opposite state, is painful."

We tend to feel pleased—or at least, relieved—when we accomplish something we want. Thus, as we remarked in Chapter 2:

When Carol recognized that her goal was achieved, she felt satisfaction, fulfillment, and pleasure—and those feelings then helped
her to learn and remember.

Of course, we're delighted that Carol felt pleased, but how did those feelings help her to learn—and why do we like those feelings so much, and
work so hard to find ways to attain them? Indeed, what does it mean to say
that someone feels "pleased"? When people answer questions like these,
we frequently hear examples of circular reasoning:

Citizen: I do the things that I like to do because I get pleasure
from doing them. And naturally, I find them pleasant because
those are the things that I like to do.

One reason why we get into such circles is that we usually cannot describe
any feeling itself, but can resort only to analogies, like *"That pain was as
piercing as a knife."* What could make something so hard to describe that
we're forced to refer to comparisons? Clearly, this is likely to happen when
we don't have a way to divide that thing—be it an object, a process, or a
mental state—into several parts, or layers, or phases. *This is because a thing
that we cannot split into parts gives us nothing to use as pieces of explanation!*
However, this goes against the popular view that such feelings as pleasure
or pain are "basic" or "elemental" in the sense that they can't be explained
in terms of these things.

However, this section will argue that what we call "pleasure" is a suitcase-word for quite a few different processes that we don't often recognize,
and this has been an obstacle to understanding our psychology. So let's
try to catalog some of the feelings and activities that make the concept of
"pleasure" more complex than it might seem.

*Satisfaction.* A species of pleasure called "satisfaction" comes when
an ambition has been achieved.
*Exploration.* We may also feel pleasure during a quest—and not
only at the end of it. So it is not only a matter of being rewarded
for achieving a goal.
*Goal-suppression.* You may also regard your condition as pleasant if
some other process has suppressed most of your other critics and
goals.
*Relief.* A species of pleasure called "relief "may come when a problem
has been solved—if that goal was represented as an irritation.

Success can also fill you with pleasure and pride—and may also motivate you to show other persons what you have done. But the pleasure of success soon fades because, no sooner is one problem put to rest than another one quickly replaces it. Besides, few of our problems stand by themselves; they are usually only parts of larger ones.

Also, after you've solved a difficult problem, you may feel relieved and satisfied, and sometimes may also feel a need to arrange for some sort of inner or outer celebration. Why might we have such rituals? Perhaps there's a special kind of relief that comes when one can dismiss a goal and release the resources that it engaged—along with the stresses that came with them. Clearing out one's mental house may help to make other things easier—just as the "closure" of a funeral can help to assuage a person's grief.

But what if the problem you're facing persists? You can sometimes regard your present distress as a benefit, as in *"I'm certainly learning a lot from this,"* or *"Others may learn from my mistakes."* And everyone knows this magical trick for turning all failures into success: one can always tell oneself, *"The true reward is the journey itself."*

So instead of trying to say what Pleasure *is,* we'll need to develop more ideas about what processes might be involved in what we often describe in simple terms—such as "feeling good." In particular, it seems to me that we often use words like *pleasure* and *satisfaction* to refer to an extensive network of processes that we do not yet understand, and when anything seems so complex that we can't grasp it all at once, then we tend to treat it as though it were single and indivisible.

> "Pleasures are ever in our hands or eyes,
> And when in act they cease, in prospect, rise:
> Present to grasp, and future still to find,
> The whole employ of body and of mind."
> —*Alexander Pope, in* Essay on Man

## The Pleasure of Exploration

> St. Augustine: "Pleasure pursues objects that are beautiful, melodious, fragrant, savory, soft. But curiosity, seeking new experiences, will even seek out the contrary of these, not to experience the discomfort that may come with them, but from a passion for experimenting and knowledge."

Understanding a new and difficult subject—or exploring an unfamiliar terrain—can lead to a lot of pain and stress. Then how can we keep this from holding us back from learning new ways to accomplish things? One antidote for this is *adventurousness*.

> Minsky 1986: "Why do children enjoy the rides in amusement parks, knowing that they will be scared, even sick? Why do explorers endure discomfort and pain—knowing that their very purpose will disperse once they arrive? And what makes people work for years at jobs they hate, so that someday they will be able to—they seem to have forgotten what! It is the same for solving difficult problems, or climbing freezing mountain peaks, or playing pipe organs with one's feet: some parts of the mind find these horrible, while other parts enjoy forcing those first parts to work for them."

Most of our everyday learning involves only minor adjustments to skills that we already know how to use. One can do this by using "trial and error"; one makes a small change, and if that results in a pleasant reward (such as being pleased with an improved performance) then that change will become more permanent. This fact has led many teachers to recommend that "learning environments" should mainly consist of situations in which pupils get frequent rewards for success. To promote this, then, it is often suggested that one should help the students to progress through a sequence of small, easy steps.

> Edward L. Thorndike 1911: "The Law of Effect is that: Of several responses made to the same situation, those which are accompanied or closely followed by satisfaction to the animal will, other things being equal, be more firmly connected with the situation, so that, when it recurs, they will be more likely to recur; those which are accompanied or closely followed by discomfort to the animal will, other things being equal, have their connections with that situation weakened, so that, when it recurs, they will be less likely to occur. The greater the satisfaction or discomfort, the greater the strengthening or weakening of the bond."

However, this pleasant and positive strategy may not work well in unfamiliar realms because when we are learning a new technique, we need to work harder with fewer rewards, while enduring the additional stress of being confused and disoriented. Also, it may require us to abandon older techniques and representations that have previously served us well—which might even arouse a sense of loss that brings "negative" feelings akin to grief. Such periods of awkwardness and ineptitude would usually cause a person to quit.

Thus "pleasant" or "positive" practice, alone, may not suffice for us to learn more radically different Ways to Think. This, in turn, suggests that to become proficient at learning new things, a person must somehow acquire what St. Augustine called, in the extract on page 323, "a passion for experimenting and knowledge." Such persons must somehow have managed to train themselves actually to enjoy those discomforts.

> Citizen: How can you speak of "enjoying" discomfort? Isn't that a self-contradiction?

It is only a contradiction when you regard your Self as a single thing. But when you see the mind as a cloud of conflicting resources, then you no longer need to think of pleasure as a "basic" or all-or-none thing. For now you can imagine that, while *some parts of your mind are uncomfortable, other parts of your mind may enjoy forcing those first parts to work for them.* For example, one part of your mind can still represent your state in a positive way, by saying, *"Good, this is a chance to experience awkwardness and to discover new kinds of mistakes!"*

> Citizen: But wouldn't you still be feeling that pain?

Indeed, when struggling at their seemingly punishing tasks, athletes still feel physical pain, and artists and scientists feel mental pains—but, somehow, they seem to have trained themselves to keep those pains from spiraling into the awful cascades we call "suffering." But how could those persons have learned to suppress, ignore, or enjoy those pains, while preventing those disruption cascades? To answer that, we would need to know more about our mental machinery.

> Scientist: Perhaps this does not really need any special explanation, because explorations can provide their own rewards. For me, few things bring more pleasure than making radical new hypotheses—

and then showing that their predictions are correct, despite the objections of my competitors.

Artist: It seems almost the same to me, because nothing surpasses the thrill of conceiving a new kind of method or representation and then confirming that this will produce new effects in my audience.

Psychologist: Many achievers regard their ability to function in spite of pain, rejection, or adversity to be among their outstanding accomplishments!

All this suggests that "exploration pleasure" (however it works) may be indispensable to those who want to keep on extending their abilities. To be sure, we usually see pleasure as positive, but one can see it as negative—because of how it tends to suppress other competing activities. More generally, to accomplish any major goal, one may need to suppress most competing goals, as in, *I don't feel like doing anything else.* Most traditional theories of learning assumed that an action that led to pleasure would be reinforced, so that you'll be more likely to react that way in the future. However, I suspect that pleasure also helps us to learn by engaging another, more "negative" function that works to keep our minds from "changing the subject" while credit assignment is being accomplished!

## 9-5 What Makes Feelings So Hard to Describe?

"A color stands abroad
On solitary hills
That science cannot overtake
But human nature feels"
—*Emily Dickinson*[8]

Many thinkers have wondered about the relations between our minds and our brains. If the bodies (of which our brains are parts) consist of nothing more than physical stuff, then each person must be some sort of machine. Of course, that machine is immensely complex; in every human embryo, billions of units of DNA are involved with assembling countless atoms and molecules into intricate arrangements of thousands of types of membranes, fibers, pumps, and pipes. Nevertheless, one still has to ask how any such structures could ever support what we call our sensations and thoughts?

> Dualist philosopher: Computers can do only what they're programmed to do, simply proceeding from step to step, without any sense that they're doing this. Machines can have no goals or aversions or pleasures or pains—or any sensations or feelings at all because they lack certain vital ingredients that can only exist in living things.

But what could those "vital ingredients" be? Many philosophers have wondered how a thing composed only of physical parts could ever "really" feel or think.

> David Chalmers 1995b: "When we visually perceive the world, we do not just process information; we have a subjective experience of color, shape, and depth. We have experiences associated with other senses (think of auditory experiences of music, or the ineffable nature of smell experiences), with bodily sensations (e.g., pains, tickles, and orgasms), with mental imagery (e.g., the colored shapes that appear when one rubs one's eyes), with emotion (the sparkle of happiness, the intensity of anger, the weight of despair), and with the stream of conscious thought.
>
> "[That we have a sense of experiencing] is the central fact about the mind, but it is also the most mysterious. Why should a physical system, no matter how complex and well-organized, give rise to experience at all? Why is it that all this processing does not go on "in the dark," without any subjective quality? Right now, nobody has good answers to these questions. This is the phenomenon that makes consciousness a *real* mystery."

However, it seems to me that the mysteries that Chalmers sees result from squeezing multiple mental activities into suitcase-words like *subjective, sensations,* and *consciousness.* For example, Chapter 4-2 showed how people use the word *consciousness* for at least a dozen mental processes—and Chapter 5-7 showed that our perceptual systems also involve many types and levels of processing. However, our higher-level processes cannot detect all those intermediate steps—and this lack of insight leads us to the belief that our sensations come to us in some way that is simple, direct, and immediate.[9]

For example, whenever something touches your hand, it seems to you that you instantly sense that you have felt a touch on your hand—*and that this happened immediately, without any complex processing.* Similarly, when you look at a color and sense that it's red, no intermediate steps seem to intervene—and so you can find nothing to say about it. Surely this is at least partly why so many philosophical thinkers conclude that there can be no "mechanical" explanation of why different stimuli seem each to have particular qualities: they simply have not worked hard enough to imagine adequate models of those processes; instead, they mainly attempted to show that no such models would ever be possible.

Now, although we find it hard to speak about the character of any particular, single, sensation, we find it far easier to compare or contrast two different but similar kinds of sensations. For example, one can say that sunlight is brighter than candlelight, or that pink lies in between red and white, or that a touch on your cheek is somewhere between your ear and your chin.

However, this says nothing about how each separate sensation "feels." It's like describing the distance between two towns on a map, while saying nothing about those individual towns. Similarly, if I were to ask what the color red means to you, you might first say that it makes you think of a rose, which then reminds you of being in love—and then you'll find yourself relating this to other kinds of sensations and feelings; red might also remind you of blood, and make you feel some sense of dread or fear. Similarly, green might make one think about pastoral scenes and blue might suggest the sky or the sea. Thus, a seemingly simple stimulus can lead to many other kinds of mental events, such as these other feelings and reminiscences.

Similarly, when you try to describe the feelings that come with being in love, or from suffering fear, or when seeing a pasture or a sea, you'll soon find that you are merely mentioning yet other things that these *remind* you of. And then, perhaps, you will come to suspect that one can never really describe what anything *is;* one can only describe what that thing is *like*.

What would be a useful alternative to the idea that our sense of "experiencing" is mysterious? Well, if your higher cognitive levels had better access to your lower ones, then you might be able to replace statements like *"I am experiencing the sensation of seeing something red"* by more detailed descriptions of the processing that sensations involve, such as

> "My resources have classified certain stimuli, and then made some representations of my situation, and then some of my Critics changed certain plans I had made, and altered some ways in which I was perceiving things, and this led to the following sorts of cascades, and so forth."

If we were able to make such descriptions, the mystery of "subjective experience" should disappear, because then we would have enough ingredients to answer our questions about those processes. In other words, it seems to me, the apparent "directness of experience" is an illusion that comes because our higher mental levels have such limited access to the systems we use to recognize, represent, and react to our external and internal conditions.

I don't mean to suggest that this illusion is usually harmful, or that we should strive to surmount all those limitations, because, as we noted in Chapter 4-4, too much such information might overload our minds; however, some such therapy might benefit some of those dualist philosophers. Also, in some future time, we will have to make decisions about the extent to which our future Artificial Intelligence machines should be equipped with ways to inspect (and then, to also be able to change) their own systems—or whether we'll need to prohibit that access.

## How Do You Know When You're Feeling a Pain?

Common sense might answer that you can't have a pain without knowing it. However, some thinkers disagree with that:

> Gilbert Ryle 1949: "A walker engaged in a heated dispute may be unconscious of the sensations in his blistered heel, and the reader of these words was, when he began this sentence, probably unconscious of the muscular and skin sensations in the back of his neck or his left knee. A person may also be unconscious or unaware that he is frowning, beating time to the music, or muttering."

Similarly, Joan might first notice a change in her gait, and only later notice that she's been favoring her injured knee. Indeed, her friends may be more aware than she is of how much that pain is affecting her. Thus, one's first awareness of being in pain may come only after detecting other signs of its effects, such as discomfort or ineffectiveness—perhaps by using the kind of machinery that we described in Chapter 4-3.

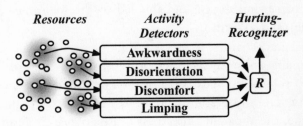

If you think you feel pain, could you be mistaken? Some would insist that this cannot be because *pain* is the same as *feeling pain*—but again, our philosopher disagrees:

> Gilbert Ryle 1949: "The fact that a person takes heed of his organic sensations does not entail that he is exempt from error about them. He can make mistakes about their causes and he can make mistakes about their locations. Furthermore, he can make mistakes about whether they are real or fancied, as hypochondriacs do."

We can make such mistakes because what we "perceive" does not come directly from physical sensors but from our higher-level processes. Thus, at first the source of your pain may seem vague because you have only noticed that something's disrupting your train of thought; then the best that you can say might be, *"I don't feel quite right, but I don't quite know why. It could be a headache just starting to hurt. Or maybe the start of a bellyache."* Similarly, when you are falling asleep, the first things you notice might be that you've started to yawn, or you keep nodding your head, or you're making a lot of grammatical errors; indeed, your friends might notice these before you do. One might even see this as evidence that people have no special ways to recognize their own mental states, but do this with the same methods they use to recognize how other persons feel.

Charles: Surely that view is too extreme. Like anyone else, I can observe my behavior "objectively." However, I also have an ability—which philosophers call "privileged access"—with which I can inspect my own mind "subjectively" in ways that no other person can.

We certainly each have some privileged access, but we should not overrate its significance. I suspect that our access to our own thoughts provides more *quantity* but does not seem to reveal much more about the nature of our own mental activities. Indeed, our self-assessments are sometimes so inept that our friends may have better ideas about how we think.

Joan: Still, one thing is sure: none of my friends can *feel* my pain. I surely have privileged access to *that*.

It is true that the nerves from your knee to your brain convey signals that none of your friends can receive. But it's almost the same when you talk to a friend through a telephone. "Privileged access" does not imply magic; it's merely a matter of privacy—and no matter how private those lines may be, you still must use other processes to assign any other significance to the signals that get to your brain from your knee. That's why Joan might find herself wondering, *"Is this the same pain that I felt last winter, when my ski boot did not release quickly enough?"*

Joan: I'm not even sure that it was the same knee. But isn't something missing here? If sensations are nothing but signals on nerves, then why are there such distinctive differences between the tastes of sour and sweet, or between the colors of red and blue?

## 9-6  The Sense of Having an Experience

William James 1890: "It is astonishing what havoc is wrought in psychology by admitting at the outset apparently innocent suppositions, that nevertheless contain a flaw. . . . The notion that sensations, being the simplest things, are the first things to take up in psychology is one of these suppositions. The only thing which psychology has a right to postulate at the outset is the fact of thinking itself, and that must first be taken up and analyzed. If

sensations then prove to be amongst the elements of the thinking,
we shall be no worse off as respects them than if we had taken
them for granted at the start."

Many philosophers have maintained that our sensations have certain
"basic" qualities that cannot be reduced to anything else. For example,
they claim that each color like red and each flavor like sweet has its own
unique "quality" that cannot be described in terms of other things.

Of course, it is not hard to make a physical instrument to measure the
amount of red light that comes from the surface of some particular apple,
or to measure the weight of the sugar contained in the flesh of any particular peach. However (those philosophers claim), such measurements tell
you nothing about the *experience* of seeing a redness or tasting a sweetness.
And then (some philosophers go on to claim), if those "subjective experiences" cannot be detected by physical instruments, they must exist in a
separate mental world, which would mean that we cannot explain how
minds work in terms of machinery inside our brains.

However, there is a serious flaw in that argument. For if you can say,
*"This apple looks red to me,"* then some "physical instrument" in your brain
must have recognized the activity involved with that experience—and then
caused your vocal tract to behave accordingly. That "experience-detecting"
instrument could be another internal activity recognizer like those we've
seen in Chapter 4-3.

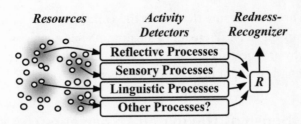

Our brain scientists have not yet located such circuits inside our
brains—but it surely is only a matter of time before we find clusters of
brain cells that recognize such combinations of conditions. Then we'll be
able to take William James's advice and start to develop more constructive
theories about the processes that we call "sensations" and "feelings."

In any case, we already know that our perceptions are far from direct.
For example, when a ray of light strikes the back of your eye, a signal will

flow from each retinal cell that this excites—and those signals will then affect other resources inside your brain—and some of those resources will then construct descriptions and reports that influence yet other parts of your brain.[10] At the same time, other streams of information will also affect those descriptions so that, when you try to describe your "experience," you'll be telling a story based on sixth-hand reports.

The idea that sensations are "basic" may have been useful in older times, but today we need to recognize the extent to which our perceptions are affected by what our other resources may want or expect. In fact, as we mentioned in Chapter 5-7, more signals flow downward to the brain's sensory cortex than in the opposite direction, presumably to help us see what we expect to see—by priming us with an appropriate "simulus." This could help to explain, for example, how we frequently "see" things that do not exist—such as the "square" below.[11]

Once we appreciate the complexity of our perceptual machinery, we can finally answer that question about why we find feelings so hard to describe. For what would a person need to be able to express their "subjective feelings"? Perhaps it is no accident that one meaning of the word *express* is "to squeeze"—for when you try to "express yourself," your language resources will have to pick and choose among the descriptions your other resources construct—and then attempt to squeeze a few of these through your tiny channels of phrases and gestures.

Of course one can never describe one's whole state of mind, because one can focus on just a few things at a time, and because one's state is constantly changing—so, usually, you will simply settle for expressing those aspects whose signals seem most urgent at each moment. At one moment you're thinking about your foot; then some other sensation attracts your attention; perhaps you notice a change in some sound, or turn your head toward something in motion—*and then you notice that you are noticing these.* So you can never be "wholly aware of yourself" because "you" are a river of rivaling interests, always enmeshed in cascades of attempts to describe its ever-changing eddies and tides.

## 9-7  How Is a Human Mind Organized?

> Jean Piaget 1923: "If children fail to understand one another, it is
> because they think they understand one another. . . . The explainer
> believes from the start that the listener will grasp everything, will
> almost know beforehand all that should be known. . . . These
> habits of thought account, in the first place, for the remarkable
> lack of precision in childish style."

How do human minds develop? We know that our infants are already
equipped at birth with ways to react to certain kinds of sounds and smells,
to certain patterns of darkness and light, and to various tactile and haptic
sensations. Then over the following months and years, the child proceeds
through many stages of mental development. Eventually, each normal
child learns to recognize, represent, and reflect upon some of his own
internal states, and also comes to self-reflect on some of his intentions and
feelings—and eventually learns to identify these with aspects of how other
persons behave.

What kinds of structures might we use to support those kinds of activities? Our previous chapters showed several views of how a human mind
might be organized. We began by portraying the mind (or brain) as being
based on a scheme that deals with various situations by activating certain
sets of resources—so that each such selection will function as a somewhat
different Way to Think.

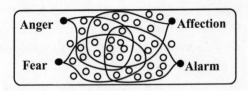

To determine which set of resources to select, such a system could
begin with simple sorts of *If→Do* rules and could later start to replace
these with more versatile *Critic→Selector* schemes.

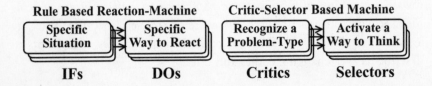

Chapters 5, 6, and 7 conjectured that the adult mind comes to have multiple levels, each of which contains additional Critics, Selectors, and other sources. We also noted that these ideas seem consistent with Sigmund Freud's early view of the mind as a system for dealing with conflicts between our instinctive and acquired ideas.

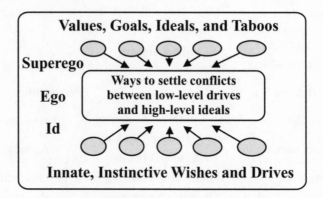

THE FREUDIAN SANDWICH

Finally, Chapter 8 went on to suggest that our various ways to represent knowledge and skills might also be arranged in levels that have increasing symbolic expressiveness.

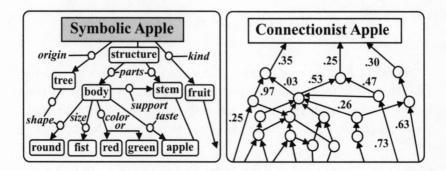

Each of those ways to envision a mind has different kinds of virtues and faults, so rather than ask which model is best, one needs to develop Critics that learn to choose when and how to use each of those models. However, none of the models that we have discussed does well to represent the organization of an entire human mind; each of them only helps us to think about certain kinds of mental activities.

In any case, we also will need a model with room enough to have places to put answers to questions that we have not even yet thought of asking. For this I have found it useful to think of the mind as though it were a decentralized cloud of yet unimagined processes, interacting in still unspecified ways. For example, one could envision something like the cloud of resources in Chapter 1, except filled with higher-level systems like those shown here.

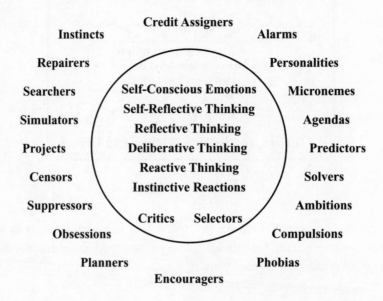

## Is a Mind Like a Human Community?

It might be tempting to portray our mental processes as organized like a typical human community—such as a residential village or town, or an industrial company. In a typical corporate organization, the human resources are arranged in accord with some formal hierarchical plan.

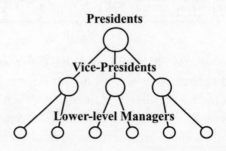

We tend to invent this kind of "management tree" whenever there's more work than one person can do; then the work is divided into parts, which are assigned to subordinates. This picture might tempt us to identify a person's Self with the chief executive of a company, who controls a "chain of command" that branches down and out.

However, this is not a good model for human brains, because an employee of a company might be able to learn to do almost any new task—whereas most parts of a brain are too specialized for that. Also, when a company becomes wealthy enough, it can expand to do new activities, by hiring additional employee minds.[12] In contrast, people do not (yet) have practical ways to expand their individual brains. In fact, it is almost the opposite: whenever you try to do several tasks at once, each of your subprocesses is likely to encounter new handicaps. Perhaps we should state this as a general principle.

*The Parallel Paradox: If you break a large job into several parts and try to work on them all at once, then each process may lose some competence, from lacking access to resources it needs.*

There is a popular belief that the brain gets much of its power and speed because it can do many things in parallel. Indeed, it is clear that some of our sensory, motor, and other systems can do many things simultaneously. However, it also seems clear that when we tackle more difficult problems, we increasingly need to divide those problems into parts, and focus on these sequentially. This means that our higher, reflective levels of thought will tend to operate more serially. This may also partly account for our sense of having (or being) a "stream of consciousness."

In contrast, when a company divides a job into parts, it can often pass them down to separate subordinates who can work more simultaneously. However, that leads to a different kind of cost:

*The Pinnacle Paradox: As an organization grows more complex, its chief executive will understand it less, and increasingly will need to place more trust in decisions made by subordinates.*

Of course, many human communities are less hierarchical than our companies are, and they use more cooperation, consensus, and compromise in making decisions and settling arguments. Such negotiations can be more versatile than either dictatorship or "majority rule" (which gives each participant a spurious sense of "making a difference," despite the fact most differences thereby get canceled out). This also raises questions about the extent to which our human "subpersonalities" can learn to cooperate to help with larger jobs.

## Central and Peripheral Controls

Every higher animal has evolved many resources—let's call them "alarmers"—that can react to certain states of affairs by interrupting higher-level processes. These conditions include reactions to such signs of possible dangers as rapid motions and loud sounds, unexpected touches, and the sightings of insects, spiders, and snakes. We also react with alarm to aches and pains, feelings of illness, and hunger and thirst. Similarly, we are subject to more pleasant kinds of interruptions, such as the sights and smells of foods to eat, and of signals of sexual interest.

Many such reactions work without interrupting your other mental activities—as when your hand moves to rub an insect bite, or your eyes turn away from excessive light. Other instinctive alarms may get more attention, such as impending collisions, extreme heat or cold, losing one's balance, loud noises or growls, or seeing a spider or a snake.

We are also subject to alarms that seem to come from "inside the mind"—as when we detect an unexpected new idea or a mental process failing to work, or a conflict among our goals and ideals, or when we react to internal conditions such as shame or surprise. A Critic-Selector Model of Mind could account for many such mental reactions, using Correctors, Censors, and Suppressors.

However, one also could use a less centralized view in which our thinking consists of interactions among many partly autonomous processes. For example, one could think of one's mind as like a city or town whose processes consist of the activities of subdepartments concerned with transportation, water, power, fire, police, school, planning, housing, parks, and streets—as well as legal and social services, public works, pest control, and so on, each with its own subadministrations.

Should one think of a city as having a Self? Some observers might argue that each town has a certain "ambience" or "atmosphere," and certain traits and characteristics. But few would insist that a city or town has anything that resembles a human mind.

Citizen: Perhaps that's because they don't have your idea that a Self is a network of various models, each of which may help a mind to answer certain questions about itself. But in fact, each of those departments for planning, power, parks, and streets has plenty of diagrams and charts that represent aspects of the town they're in.

Programmer: Some modern computer systems work by combining multiple processes, each monitoring some of the others—but it is hard to make such systems work reliably. So I wonder how all your resources could be combined to work together dependably. What happens if some of its parts break down? A single error in a large computer program can cause the entire system to stop.

I suspect our human "thinking processes" often "break down," but you rarely notice that anything's wrong, because your systems so quickly switch you to think in different ways, while the systems that failed are repaired or replaced. Here are a few of the kinds of failures that are likely to get more "attention."

You have trouble recalling past events.
You have trouble when solving an urgent problem.
You cannot decide which action to take.
You've lost track of what you were trying to do.
Something has happened that surprises you.

Nevertheless, in cases like these, you can usually switch to other tactics and strategies. For example, you might change the domain you are searching through, or select some other problem to solve, or switch to some different overall plan, or make a major change in emotional state—without noticing that you are doing this.

Furthermore, whenever some of your systems fail, your brain may retain some earlier versions of them. Then in situations where you get confused, you may be able to ask yourself, *"How did I deal with such things in the past?"* and this might cause some parts of your mind to "regress" to an earlier version of themselves, from an age when such matters seemed simpler to you. This suggests another reason as to why we might like the idea of having a Self:

Minsky 1986: "One's present personality cannot share all the thoughts of one's older personalities—and yet it has some sense that they exist. This is one reason why we feel that we possess an inner Self—a sort of ever-present person-friend, inside the mind, whom we can always ask for help."

However, we should not ignore the fact that people are also subject to failures from which recovery may be difficult or impossible. For example, if something went wrong with the machinery that controls your Critic/Selector processes, then the rest of your mind may become reduced to a disorganized cloud of conflicting resources—or get stuck with some single, unswitchable way to think. Here, again, is Lovecraft's observation:

H. P. Lovecraft 1926: "The most merciful thing in the world, I think, is the inability of the human mind to correlate all its contents. We live on a placid island of ignorance in the midst of black seas of infinity, and it was not meant that we should voyage far. The sciences, each straining in its own direction, have hitherto harmed us little; but some day the piecing together of dissociated knowledge will open up such terrifying vistas of reality, and of our frightful position therein, that we shall either go mad from the revelation or flee from the deadly light into the peace and safety of a new dark age."

## Mental Bugs and Parasites

It seems safe to predict that most of our future attempts to build large, growing Artificial Intelligences will be subject to all sorts of mental disorders. For if a mind could make changes in how it works, it would face the risk of destroying itself. This could be one reason why our brains evolved so many partly separate systems, instead of a more unified and centralized one: there may have been substantial advantages to imposing limits on the extent to which our minds could examine themselves!

For example, no single way to think should be permitted to gain too much control over the systems we use for remembering. For then it might be able to overwrite all that person's old memories. Similarly it could be dangerous for any resource to remain very active for long, because then it might force the rest of the mind to spend all its time pursuing one particular goal. Also, if any resource were able to completely suppress some instinctive drives, then that resource might be able to force its person to never sleep, or to work to death, or to starve itself—and the same would apply to any resource that could control our systems for pleasure and pain.

While such drastic calamities are rare, many common human faults result from the growth of "mental parasites" that take the form of the self-reproducing sets of ideas that Richard Dawkins called "memes." Such a collection of concepts may include ways to grow and protect itself by displacing competing sets of ideas; see Susan Blakemore 1999. To protect themselves from such extremes, our brains evolved ways to balance between becoming too highly centralized or too dispersed to have much use. We had to be able to concentrate, yet also to respond to urgent alarms. Nevertheless, we still remain prone to doctrines, philosophies, faiths, and beliefs that spread through the populations of entire civilizations. It is hard to imagine any foolproof ways to protect ourselves from such infections. So far as I can see, the best we can do is to try to educate our children to learn more skills of critical thinking and methods of scientific verification.

## 9-8 The Dignity of Complexity

> Vitalist: Your theories are too mechanical, they're filled with parts but have no wholes. What they need is some kind of lifelike cement—some coherent essence to hold them together.

I sympathize with that quest for cement, but I can't see how it could actually help, because whatever adhesive you might propose—such as a single central Self—you'd still be obliged to describe *its* parts, and the magical glue that binds them together. So words like *spirit* or *essence* serve only to make us keep asking the very same questions. As for terms like *me, myself,* and *I,* these seem only to describe the times when we're using our models of our minds.

> Citizen: But surely, no being would want to think of itself as a disorderly mess of contraptions and gadgets. How could it have any self-respect when it sees itself as nothing more than the product of countless thousands of accidents?

Perhaps the most popular concept of what we are assumes that we each have a central core—some sort of invisible spirit or ghost that comes to us as an anonymous gift. However, a more realistic view would recognize that each human mind that exists today is one result of a process in which decillions of previous creatures on Earth spent their lives reacting, adjusting, adapting, and dying so that some of their descendants might thrive. In that unthinkably vast history, all those creatures contributed to a tremendous set of experiments, each of which may have contributed to giving us slightly more powerful brains.

We don't yet know how that process began, except that it probably started in some small watery place and spread out to inhabit the oceans, seashores, deserts, and plains until our ancestors developed ways to live in the villages, cities, and towns that they built. However, we do know that this struggle went on for thirty million centuries and—so far as we know—no other magnificent process like this has ever occurred in our universe. Yet most traditional accounts of our origins make no mention whatsoever about this prodigious saga of sacrifice.

All this suggests that it would be reckless, rash, and negligent to dismiss our mental abilities as though they came as gratuitous gifts. For until we find good evidence for other intelligence in this universe, we should recognize how much we owe to all those creatures who died for us—and take care to ensure that the minds we inherit don't all go to waste from our making some foolish, world-wrecking mistake.

## 9-9 Some Sources of Human Resourcefulness

Our unrivaled human resourcefulness developed over three vastly different scales of time.

> Genetic Endowment: The genes that shape our modern brains were selected from variations that occurred during the last half billion years.

Each human brain contains hundred of different resources, each composed of millions of clusters of cells that come in many different varieties. These inherited systems help us to escape from various kinds of dangers and threats.

> Cultural Heritage: The sets of beliefs in each culture evolved over many centuries in which that human community selected ideas produced by its individuals.

Our cultural traditions are the principal sources of every citizen's knowledge and skills because no person, alone, could ever invent as many ideas as a four-year-old child learns.

> Individual Experience: Each year, one learns millions of fragments of knowledge from our own private experiences.

For example, consider how much knowledge hides in virtually every language word; if you listen to anything anyone says, you'll hear many useful analogies. We speak of time as though it resembles space—as when a listener wonders when a speaker will *get* to some *point*. Also, we often think of time as a fluid that's "running out," and we talk about friendships in physical terms, as in *"Carol and Joan are very close."* All of our language is riddled and stitched with cross-linked ways of portraying things—and sometimes we call these "metaphors." (See Lakoff 1980.)

Some metaphors seem quite pedestrian, such as when we speak of "taking steps" to cause or prevent some happening. Other metaphors seem more remarkable—such as when a scientist thinks about a fluid as though it were like a bundle of tubes. We notice such analogies when they play surprisingly productive roles, but we rarely observe how frequently we use the same techniques in commonsense thinking. (See Lakoff 1980.)

Some analogies have simple origins, as when they come from stripping away enough details to make two different subjects seem the same. More complex metaphors represent things as though they existed in other realms in which we can use other familiar skills. I suspect that most of our commonsense knowledge may be embodied as metaphors in the forms that Chapter 6 called "panalogies."

How do we learn those precious panalogies? I suspect that some of them (such as those analogies between space and time) are virtually born in our brains, because certain regions of our brains are genetically wired so that we can scarcely help but represent different realms of ideas as having analogous properties. However, on some occasions, some individual will discover a new kind of description or representation or formulation that is both so fruitful and so easy to use that it spreads through that person's community. Naturally, we'd like to know how those fertile discoveries were made, but many of those rare events may never be explained at all—because, like the mutations of our genes, such events need to happen only once, and then can spread from brain to brain. However, other pervasive analogies may be so "natural" and inevitable that almost every child, by itself, will invent the same ones without any help.

All of these kinds of inventiveness, combined with our unique expressiveness, have empowered our communities to deal with huge classes of new situations. The previous chapters discussed many aspects of what gives people so much resourcefulness: each of our processes has deficiencies, but we can usually find alternatives.

We have multiple descriptions of things—and can quickly switch among them.

We make memory-records of what we've done—so that later we can reflect on them.

Whenever one of our Ways to Think fails, we can switch to another.

We split hard problems into smaller parts, and keep track of them with our context stacks.

We manage to control our minds with all sorts of bribes, incentives, and threats.

We have many different ways to learn and can also learn new ways to learn.

Nevertheless, our minds still have bugs. For, as our human brains evolved, each short-term advance exposed us to the danger of making new types of mistakes. For example, our wonderful powers to make abstractions can lead us to overlook vital exceptions. Our capacious memory systems are likely to accumulate wrong or misleading information. Our juvenile attachment-learning schemes often lead us to believe whatever our Imprimers believe. Our imaginations are so powerful that we confuse realities with our fantasies; then we get obsessed with unachievable goals, and set out on extensive but futile quests—or become so loath to accept a failure or loss that we try to regress to our lives of the past.

We cannot expect to escape from all such bugs because, as every engineer knows, most every change in a complex system will introduce yet other troubles that won't show up till the system moves to a different environment. Also, each human brain is unique because it is built by pairs of inherited genes, each chosen at random from one of its parents; then, in that brain's early development, many smaller details will depend on other accidental events. How could such machines work reliably, in spite of so much variety? To explain this, quite a few thinkers have argued that our brains must be based on "holistic" principles, according to which every fragment of process or knowledge is "globally distributed" (in some unknown way) so that the system's behavior would still be the same in spite of the loss of some of its parts.

However, the arguments in this book suggest that we do not need any such magical tricks—because we have so many different ways to accomplish any type of job. Also, it makes sense to suppose that many parts of our brains evolved as ways to correct (or to suppress) the effects of defects in other parts. This means that our scientists will find it hard to discover how human brains manage to work so well, and why they evolved in the ways that they did. I suspect that we won't understand such things until we have had more experience at trying to build such systems ourselves. Only then will we learn enough about which kinds of bugs we're likely to find and about how to keep them under control.

In the coming decades, many researchers will try to develop more resourceful machines, but every system that we build will surprise us with new kinds of flaws—until those machines become clever enough to conceal their faults from us. Sometimes we'll be able to diagnose specific errors in those designs and then we'll be able to remedy the errors. But whenever we fail to find ways to make such repairs, we will have little choice except to add more checks and balances—for example, by adding more Critics and Censors. However, we'll never find any one, foolproof way to choose (for example) between the advantages of immediate actions and the benefits of cautious, reflective thinking. So whatever we do, we can be sure that the road toward "posthuman minds" won't be smooth.

# ACKNOWLEDGMENTS

This book develops some new ideas about our human mental activities. In some ways this is a sequel to my earlier book, *The Society of Mind*, which suggested some theories of how thinking works, when seen as a collection of partly separate processes. In contrast, *The Emotion Machine* is mainly concerned with our highest and most reflective thoughts, so that there is not much overlap between the subjects that these two books discuss. To understand any complex thing, one needs to look at it from several perspectives—and this book attempts to describe the mind as though it could be seen "from inside"—while also suggesting techniques we could use for trying to build machines that "think."

Most of these theories came mainly from me, but many related ideas have also appeared in earlier works by such researchers as Seymour Papert, Aaron Sloman, and Daniel Dennett, as well as in the work of my students. In particular, over the previous decade my student, the late Push Singh, was my most constant collaborator.

Much of my thinking in earlier years was strongly influenced by such powerful minds as Seymour Papert, John McCarthy, Warren McCulloch, Manuel Blum, Ray Solomonoff, Claude Shannon, Oliver Selfridge, Allen Newell, Herbert Simon, Roger Schank, Douglas Lenat, Edward Fredkin, and Kenneth Haase. Also, many ideas came from my wife, Gloria Rudisch, MD, during the twenty years it took to develop these theories in this book.

Some of the processes described herein have been shown to work in computer programs developed by Michael Travers, Robert Hearn, Nicholas Cassimatis, and Push Singh. However, most of the theories in this book have not yet been assembled into a larger working system, and I hope that this book will inspire future researchers to have that ambition.

Many of these concepts also evolved from discussions on Internet newsgroups—especially with Chris Malcolm, Gary Forbis, Richard Long, Mark Rosenfelder, Neil Rickert, and many others. I also would like to recognize my discussions with such colleagues as Alan Kay, Carl Sagan, Danny Hillis, Edward Feigenbaum, Edward Fredkin, Gerald Sussman, Graziella Tonfoni, Hans Moravec, Jerome Lettvin, Joel Moses, John Nash, Nicholas Negroponte, Nils Nillsson, Patrick Gunkel, Patrick Winston, Richard Dawkins, Richard Feynman, Roger Schank, Russell Kirsch, Stephen Pinker, and Woodrow Bledsoe, and too many others to recollect.

Also many improvements came from Barbara Barry, Cynthia Solomon, Danny Hillis, David Yarmush, Dean S. Edmonds, Eray Ozkural, John Nash, Lloyd Shapley, Mortimer Casson, Ray Kurzweil, and Russell Kirsch—as well as from such thoughtful writer friends as Arthur C. Clarke, David Brin, Frederick Pohl, Greg Egan, Gregory Benford, Harry Harrison, Isaac Asimov, James P. Hogan, Jerry Pournelle, Larry Niven, Robert Heinlein, and Vernor Vinge.

Finally, I want to acknowledge suggestions that came from hundreds of students over the years, including Adolfo Guzman, Alison Druin, Ben Kuipers, Carl Hewitt, Carol Srohecker, Curtis Marx, Daniel Bobrow, Daniel Gruhl, David Levitt, David MacDonald, David Waltz, Douglas Riecken, Dustin Smith, Edwina Rissland, Eugene Charniak, Eugene Freuder, Gary Drescher, Greg Gargarian, Howard Austin, Ian Eslick, Ira Goldstein, Ivan Sutherland, Jack Holloway, James Slagle, Jeremy Wertheimer, John Amuedo, Karl Sims, Kenneth Forbus, Larry Krakauer, Larry Roberts, Louis Hodes, Manuel Blum, Michael Hawley, Renata Bushko, Richard Greenblatt, Robert Lawler, Scott Fahlman, Stephen Smoliar, Steve Strassman, Terry Winograd, Thomas Evans, Tom Knight, Warren Teitelman, William Gosper, William Henneman, William Martin, and Yoichi Takebayashi. I regretfully apologize to the many others whose contributions I've overlooked mentioning here.

Many of the references and quotations were found using the collection of ebooks in Michael Hart's website Project Gutenberg.

I am also grateful for extensive help given, in editing various versions of this book, by Nancy Mindick, Monica Strauss, Gloria Rudisch, Dustin Smith, Betty Lou McClanahan, and countless students in the courses I've taught.

Also, I want to acknowledge the generous support for this work from Jeffrey Epstein, Kazuhiko Nishi, and the endowment of my MIT Professorship by the Toshiba Corporation—and to acknowledge as well the superb environment provided by Nicholas Negroponte and the MIT Media Lab.

# NOTES

## CHAPTER 1. FALLING IN LOVE

1. Adapted from Barry Took and Marty Feldman, *Round the Horne,* BBC Radio, 1966.
2. Adapted from a note by Aaron Sloman in comp.ai.philosophy, May 16, 1995.
3. Adapted from Nikolaas Tinbergen, *The Study of Instinct* (London: Oxford University Press, 1951).
4. Rebecca West, *The Strange Necessity* (New York: Doubleday, 1928).

## CHAPTER 2. ATTACHMENTS AND GOALS

1. This could relate to psychoanalytic theories about how such objects can help to make transitions from early attachments to other kinds of relationships. See, for example, www.mythosandlogos.com/Klein.htm.
2. The idea of a "meme"—a package of information passed from one mind to another—was developed in Richard Dawkins, *The Selfish Gene* (New York: Oxford University Press, 1989). See also Susan Blackmore, *The Meme Machine* (New York: Oxford University Press, 1999), and Daniel C. Dennett, *Darwin's Dangerous Idea* (New York: Simon and Schuster, 1995).
3. See John Bowlby, *Attachment* (New York: Basic Books, 1973), 217. Bowlby bases some of his discussion on the research of H. R. Schaffer and P. E. Emerson, found in "The Development of Social Attachments in Infancy," *Monographs for the Society of Research in Child Development* 29, no. 3 (1964), 1–77, who also discuss the effects of multiple attachments.
4. Here, Bowlby is referring to Y. Spencer-Booth and R. A. Hinde, *Animal Behavior* 19 (1971): 174–191, 595–605.
5. There is some evidence that infants can imitate lip and tongue protrusion, mouth opening, and finger movement. Also see Charles A. Nelson, "The Development and Neural Bases of Face Recognition," *Infant and Child Development* 10 (2001): 3–18, and Andrew N. Meltzoff and M. Keith Moore, "Explaining

Facial Imitation: A Theoretical Model," *Early Development and Parenting* 6 (1997): 179–192.

6. Jaak Panksepp's experiments (Panksepp, *Affective Neuroscience* [New York: Oxford University Press, 1998]) suggest that imprinting resembles addiction and that separation-distress may be similar to pain because both are relieved by opioids. Panksepp also discusses Howard Hoffman's [1996] conjecture that some aspects of an object's motion or shape can release endorphins in the imprintee's brain, causing the object to seem "familiar" enough to overcome fearful reactions.

7. From a 1961 letter to Mrs. H. L. Austin.

## CHAPTER 3. FROM PAIN TO SUFFERING

1. This diagram is adapted from http://www.christianhubert.com/hypertext/brain2.jpeg.

2. Larry Taylor gave permission for this quote from his unpublished essay titled "G. Gordon Liddy, Agent from CREEP."

3. This recipe summarizes some discussions in William James, *The Varieties of Religious Experience* (New York: Random House, 1994).

4. For more details of this episode, see Chapter 4-5 of Minsky, *The Society of Mind* (New York: Simon and Schuster, 1986).

## CHAPTER 4. CONSCIOUSNESS

1. I see Howard Gardner's 2002 research as a major step toward unpacking the suitcase-word *intelligence*.

2. Some psychologists use the word *subcognitive* rather than *unconscious*.

3. It is often assumed that all our knowledge is "grounded" in worldly experience. However, in Minsky, "Interior Grounding, Reflection, and Self-Consciousness" (found in *Proceedings of an International Conference on Brain, Mind and Society*, Tohoku University, Japan, September 2005), I suggest that if we think of each part of the brain as interacting only with some other brain parts connected to it, then several such parts could learn, simultaneously, some ways to deal with their local environments. See http://web.media.mit.edu/~minsky/papers/Internal%20Grounding.html.

4. Melissa Lee Phillips, "Seeing with New Sight" (available at http://faculty.washington.edu/chudler/visblind.html) describes some problems encountered by persons who have their vision restored after growing up without much visual experience. Perhaps we can see those persons as forced to work in McDermott's engine room.

5. Adapted from Chapter 6-1 of Minsky, *The Society of Mind* (New York: Simon and Schuster, 1986).

6. Our earliest robot once started to build an arch by first placing the top of the arch in what was to be its final location! It did not yet know enough to predict that an unsupported block would fall down.

7. See Chapter 25-4 of Minsky, *The Society of Mind.*
8. This idea is explained in more detail in "Matter, Mind, and Models," a chapter in Minsky, *Semantic Information Processing* (Cambridge, Mass.: MIT Press, 1968). Full text is available at http://web.media.mit.edu/~minsky/papers/Matter MindModels.txt.
9. See http://www.imprint.co.uk/online/new1.html.
10. An individual cell in the brain may have connections to thousands of other cells, but larger regions of the brain tend to have fewer connections to other such regions.

## CHAPTER 5. LEVELS OF MENTAL ACTIVITIES

1. See http://en.wikipedia.org/wiki/Computer_chess and http://en.wikipedia.org/wiki/Adriaan_de_Groot.
2. It took hundreds of millions of years to evolve the sensory systems that we use to recognize events in the outside world. However, it could have been much easier to evolve ways to recognize higher-level brain events—if those systems used simpler representations. This could be one reason why the higher levels of human thought could advance so much in only the last few million years.
3. According to Nikolaas Tinbergen, *The Study of Instinct* (London: Oxford University Press, 1951), when an animal can't make a decision, this often results in dropping both alternatives and doing something that seems to be quite irrelevant. However, these "displacement activities" seem so fixed as to suggest that those animals lack thoughtful ways to deal with such conflicts.
4. This is sometimes called Occam's razor, and is attributed to the fourteenth-century logician William of Occam.
5. Some early steps in that project are described in Marvin Minsky and Seymour Papert, *Progress Report on Artificial Intelligence* (available at http://web.media.mit.edu/~minsky/papers/PR1971.html).
6. In fact, that darker horizontal streak is *not* the lower edge; it is part of the slightly shadowed worn-down surface next to that edge.
7. This program was based on ideas of Yoshiaki Shirai and Manuel Blum. See ftp://publications.ai.mit.edu/ai-publications/pdf/AIM-263.pdf. I should note that *Builder* was able to deal only with very neat geometrical scenes, and even today there still are no general-purpose "seeing machines" that can recognize the everyday objects in a typical room. To do this, a machine would need the kinds of real-world knowledge discussed in Chapter 6.
8. See papers by Adolfo Guzman at ftp://publications.ai.mit.edu/ai-publications/pdf/AIM-139.pdf, and by David Waltz at ftp://publications.ai.mit.edu/ai-publications/pdf/AITR-271.pdf.
9. Some persons claim to imagine scenes as though looking at a photograph, whereas other persons report no such vivid experiences. However, some studies appear to show that both are equally good at recalling details of remembered scenes.

10. See, for example, http://www.usd.edu/psyc301/Rensink.htm and http://nivea .psycho.univ-paris5.fr/Mudsplash/Nature_Supp_Inf/Movies/Movie_List.html.

11. This prediction scheme appears in Section 6-7 of Minsky, "Neural-Analog Networks and the Brain-Model Problem," PhD diss., Princeton University, 1954.

## CHAPTER 6. COMMON SENSE

1. See Goethe's poem, *Der Zauberlehrling,* at http://www.fln.vcu.edu/goethe/zauber .html.

2. In a program developed by Push Singh, two robots actually consider such questions. See Push Singh, Marvin Minsky, and Ian Eslick, "Computing Commonsense," *BT Technology Journal* 22, no. 4 (October 2004) and Singh 2005a.

3. Roger C. Schank, *Conceptual Information Processing* (New York: American Elsevier, 1975), suggested some of these ideas about the meanings of the prefix *trans.*

4. See Douglas B. Lenat, *The Dimensions of Context Space,* available at http://www .cyc.com/doc/context-space.pdf. The CYC project is described at www.cyc.com.

5. This discussion is adapted from my introduction to Minsky and Seymour Papert, *Perceptrons,* 2nd edition (Cambridge: MIT Press, 1988).

6. We frequently hear tales about prodigies who have memorized huge bodies of knowledge. However, I am skeptical of such accounts, because we never see reports of experiments done to rule out deceptive performances.

7. Here the term *bit* of information is used in the technical sense that was defined in Claude E. Shannon, "A Mathematical Theory of Communication," *Bell System Technical Journal* 27 (July and October 1948). According to Ronald Rosenfeld, "A Maximum Entropy Approach to Adaptive Statistical Language Modeling," *Computer, Speech and Language* 10 (1996), the information in typical text is approximately six bits per word. If a person were to learn two bits per second for ten hours per day, then thirty years of this would come to only about one billion bits of information, which is less than the capacity of a single compact disk. See also Ralph Merkle's description at http://www.merkle.com/humanMemory.html.

8. My impression is that this also applies to the results reported in R. N. Haber, "20 Years of Haunting Eidetic Imagery: Where's the Ghost?" *Behavioral and Brain Sciences* 2 (1979), 583–629.

9. See the essays about self-organizing learning systems by Raymond J. Solomonoff: "An Inductive Inference Machine," *IRE Convention Record,* section on Information Theory, Part 2, 1957: 56–62; "A Formal Theory of Inductive Inference," *Information and Control* 7 (1964): 1–22; and "The Discovery of Algorithmic Probability," *Journal of Computer and System Sciences* 55, no. 1 (1997). See also Malcolm Pivar 1966; Douglas B. Lenat and Jon S. Brown, "Why AM and Eurisko Appear to Work," *Artificial Intelligence* 23 (1983); Douglas B. Lenat, "Eurisko: A Program Which Learns New Heuristics and Domain Concepts," *Artificial Intelligence* 21 (1983); Kenneth W. Haase, "Exploration and Invention

in Discovery," PhD diss., MIT, 1986 (text available at http://web.media.mit
.edu/~haase/thesis); Kenneth W. Haase, "Discovery Systems," found in *Advances
in Artificial Intelligence* (North-Holland: European Conference on Artificial
Intelligence, 1986); and Gary Drescher, *Made-Up Minds* (Cambridge, Mass.:
MIT Press, 1991). In recent years some of these ideas evolved into a field of
research called "Genetic Programming."

10. After a system has reached a local peak, each small change will make things worse
until one approaches a higher peak some distance away in the "fitness space."

11. At the time of this writing, some researchers are trying to "annotate" the texts
on the Web with links to the meanings of words and phrases, but I doubt that
this will work well until those networks use structures like panalogies. There
has also been recent progress toward extracting large bodies of commonsense
knowledge with the help of thousands of users of the Web. See the descriptions
of the "Open Mind Common Sense" project in Push Singh, Thomas Lin, Erik T.
Mueller, Grace Lim, Travell Perkins, and Wan Li Zhu, "Open Mind Common
Sense: Knowledge Acquisition from the General Public," found in *Proceedings
of the First International Conference on Ontologies, Databases, and Applications
of Semantics for Large Scale Information Systems,* Irvine, Calif., and at http://csc
.media.mit.edu/ and http://commonsense.media.mit.edu/.

12. This piggy-bank story was discussed extensively by Eugene Charniak in "Toward
a Model of Children's Story Comprehension," PhD diss., MIT, 1972 (also available at ftp://publications.ai.mit.edu/ai-publications/pdf/AITR-266.pdf); it has
led to some of the theories in Minsky, *A Framework for Representing Knowledge*
(Cambridge, Mass.: MIT Press, 1974), and Minsky, *The Society of Mind* (New
York: Simon and Schuster, 1986).

13. In each cycle of operation, the *General Problem Solver* program finds some differences between the current and desired states. Next, it uses a separate body of
knowledge to guess which difference is most significant, and then it tries some
methods that are likely to reduce that type of difference. Newell 1960a and Newell and Herbert A. Simon, "GPS, a Program That Simulates Human Thought,"
in *Computers and Thought,* edited by E. A. Feigenbaum and J. Feldman (New
York: McGraw-Hill, 1963), describe how, when the difference-reducing processes fail, the system will attempt to switch to a different way to represent the
situation.

14. There is no reason to assume that a system must have a central, top-level goal,
such as a "basic survival instinct," that aims toward keeping each animal alive.
Each animal has many separate instincts, such as hunger, thirst, and defense,
each of which evolved independently—but there is no reason to suppose (except,
perhaps, in the human case) that there is any representation of "being alive" anywhere in that animal's brain.

15. This could be seen as describing what programmers call a "top-down search
tree."

16. Chapter 22-10 of Minsky, *The Society of Mind,* conjectures that we use a Difference-Engine process whenever two people attempt to communicate.

17. See articles on "change-blindness": Peter Kaiser, "The Joy of Visual Perception," available at http://www.yorku.ca/eye/thejoy.htm; Kevin O'Regan, "Change-Blindness," available at http://nivea.psycho.univ-paris5.fr/ECS/ECS-CB.html; and Kevin O'Regan, "Change Blindness as a Result of Mudsplashes," in *Nature*, August 2, 1998. However, many of our sensors detect certain especially harmful conditions, and those sensors respond with signals that do not so quickly fade away.

18. Roger Schank, *Tell Me a Story* (New York: Charles Scribner's Sons, 1990), has conjectured that representing events as stories may be one of our principal ways to learn and remember.

19. See more theories about musical perception in Minsky, "Music, Mind, and Meaning," *Computer Music Journal* 5, no. 3 (Fall 1981).

20. In Clynes 1978 the musician-physiologist Manfred Clynes has described certain temporal patterns, each of which seems to help to induce a particular kind of emotional state.

21. One could ask the same questions about gossip, sports, and games. See New Zealand Time Use Study at http://www.stats.govt.nz/analytical-reports/time-use-survey.htm.

22. Full text of *The Arabian Nights* is at http://www.gutenberg.net/etext94/arabn11.txt.

23. See http://cogsci.uwaterloo.ca/Articles/Pages/how-to-decide.html.

24. Chapter 30-6 of Minsky, *The Society of Mind*, discusses why the idea of free will seems so powerful. There are many more ideas about this in Daniel C. Dennett, *Elbow Room: The Varieties of Free Will Worth Wanting* (New York: Oxford University Press, 1984).

25. See Edward A. Feigenbaum and Julian Feldman, eds., *Computers and Thought* (New York: McGraw-Hill, 1963), for more of the accomplishments of that period.

26. People sometimes use "abstract" to mean "complex" or "highly intellectual"— but here I mean almost the opposite: a more abstract description describes fewer details, which makes it applicable to more situations.

27. Papert's principle is discussed in more detail in Chapter 10-4 of Minsky, *The Society of Mind*.

## CHAPTER 7. THINKING

1. At the lowest levels, the Critics and Selectors become the same as the *If*s and *Then*s of simple reactions. At the reflective and higher levels, the Critics will tend to engage more resources and processes. Push Singh and Marvin Minsky, "An Architecture for Combining Ways to Think," in *Proceedings of the International Conference on Knowledge Intensive Multi-Agent Systems* (Cambridge, Mass.), discuss "Reflective Critics" with such abilities; and Singh, "EM-ONE: An Architecture for Reflective Commonsense Thinking," PhD diss., MIT, June 2005 (also available at http://web.media.mit.edu/~push/push-thesis.pdf), describes a working prototype of such a system, but there remains much more to do before we have a functioning six-level model.

2. Logic can be useful after a problem is solved, for justifying one's reasoning and for refining one's credit assignments; it may also be useful for making the credit assignments that we'll discuss in Chapter 8-5. There are many important discussions about the role of logic in commonsense thinking on John McCarthy's website at http://www-formal.stanford.edu/jmc/frames.html.

3. John Laird, Allen Newell, and Paul S. Rosenbloom, "Soar: An Architecture for General Intelligence," *Artificial Intelligence* 33, no. 1 (1987), describe a goal-based solving program called SOAR, which classifies obstacles into four types; Manuela Viezzer, "Ontologies and Problem-Solving Methods," 14th European Conference on Artificial Intelligence, Humboldt University, Berlin, August 2000 (also at www.cs.bham.ac.uk/~mxv/publications/onto_engineering), is a useful survey of other attempts to classify Problem Types.

4. This could be related to why some brain waves become irregular when our thinking encounters obstacles.

5. Figure 7-6 includes the names of some current ideas about how such records are represented. One can find many descriptions of these schemes by searching the Web for keywords like *sensory memory, episodic memory, short-term memory,* and *working memory,* etc. The ideas of Bernard J. Baars, "Understanding Subjectivity: Global Workspace Theory and the Resurrection of the Observing Self," *Journal of Consciousness Studies* 3, no. 3 (1996): 211–216, seem especially relevant to me.

6. The construction of long-term memories appears to involve certain phases of sleep, in ways that are not yet understood. It also appears that different kinds of memories are each stored in somewhat different ways and in different locations in the brain, such as records of autobiographical events, about other kinds of episodes, about what are called "declarative" facts, and about perceptual and motor events.

7. Chapter 19-10 of Minsky, *The Society of Mind* (New York: Simon and Schuster, 1986), described a scheme called *"Closing the Ring,"* which could help to reconnect some of the parts that were not at first retrieved.

8. This is a version of a scene described in Chapter 1 of Minsky, *The Society of Mind.*

9. See, for example, L. Friedrick-Cofer and A. C. Huston, "Television Violence and Aggression: The Debate Continues," *Psychological Bulletin* 100 (1986): 364–371.

# CHAPTER 8. RESOURCEFULNESS

1. Alan Turing, "On Computable Numbers" (available at http://www.abelard.org/turpap2/tp2-ie.asp#section-1), described these "universal" machines before any modern computers were built. The most important "small structural change" was to store the computer's program inside its rewritable memory bank so that a program could change itself, and thus was potentially able to learn; earlier computers stored programs in external devices. For a simpler description of how

these work, see Turing's "Computing Machinery and Intelligence," *Mind* 49 (1950). Subsequently, it turned out that one could build universal machines using remarkably small sets of parts.

2. This switching usually happens so quickly that we don't notice it; this is a typical instance of the Immanence Illusion of Chapter 4-5.

3. It was recently discovered that people often do not perceive some very large changes in a scene. See articles on change-blindness by Peter Kaiser, *The Joy of Visual Perception,* available at http://www.yorku.ca/eye/thejoy.htm; and by Kevin O'Regan: "Change-Blindness," available at http://nivea.psycho.univ-paris5.fr/ECS/ECS-CB.html, and "Change Blindness as a Result of Mudsplashes," in *Nature,* August 2, 1998.

4. See Chapter 3 of William H. Calvin, *How Brains Think* (New York: Basic Books, 1966).

5. For more details about changes in visual appearances, see Chapter 24 of Minsky, *The Society of Mind* (New York: Simon and Schuster, 1986), which also tries to explain why the shapes of things don't seem to change when we look at them from different directions and why things do not appear to change their locations when you move your eyes.

6. Hume was especially concerned with this question of how evidence can lead to conclusions: "It is only after a long course of uniform experiments in any kind, that we attain a firm reliance and security with regard to a particular event. Now where is that process of reasoning which, from one instance, draws a conclusion, so different from that which it infers from a hundred instances that are nowise different from that single one? I cannot find, I cannot imagine any such reasoning."

7. How could a brain compare or make copies of elaborate, network-like representations? In Chapters 22 and 23 of Minsky, *The Society of Mind,* I conjectured that this could be done only by using serial processes, and suggested that our brains use Difference-Engine techniques for making (and changing) copies of memories, as well as for communicating with verbal expressions.

8. Note that this is a Difference-Engine "in reverse"; it changes the internal description rather than changing the actual situation.

9. Some of our memory systems use certain short-lived chemicals so that those memories will quickly fade unless those chemicals keep being refreshed—whereas it appears that our longer-term memories depend on the synthesis of longer-persisting connections between the cells of the brain. Also, some information can be stored "dynamically," by being repeatedly echoed as signals inside circular loops of cells in the brain. However, Chapter 4-2 of Marvin Minsky, "Neural-Analog Networks and the Brain-Model Problem," PhD diss., Princeton University, 1954, suggests that simple such loops cannot hold much data.

10. Perhaps Carol used that facial expression to help her maintain her concentration. If this became part of her subsequent skills, it could later be hard to eliminate.

11. In the field of Artificial Intelligence, the importance of credit assignment was recognized by Arthur L. Samuel, "Some Studies in Machine Learning Using the Game of Checkers," *IBM Journal of Research and Development* 3 (July 1959):

211–219, in his early research on learning machines. Psychologists should focus more on questions about how people try to find out why and how each particular method can help to solve certain particular kinds of problems.

12. People often describe such moments as the times at which they make their decisions—and then regard these as "acts of free will." However, one might instead regard those moments as merely the times at which one's "deciding" comes to a stop.

13. Presumably, different parts of the same person's mind could use different methods for credit assignments.

14. Some of this section is adapted from Chapter 7-10 of Minsky, *The Society of Mind.*

15. For more details, see Minsky, *A Framework for Representing Knowledge;* Ross Quillian's thesis (reprinted in Marvin Minsky, ed., *Semantic Information Processing* [Cambridge, Mass.: MIT Press, 1968]); and Patrick H. Winston, ed., *The Psychology of Computer Vision* (New York: McGraw-Hill, 1975).

16. Where do we get those default assumptions? In Minsky 1974, *A Framework for Representing Knowledge,* I suggested that one usually makes a new frame by copying an older one while making some changes; then values that were not changed at that time will be inherited from those older ones.

17. This is another instance of the Immanence Illusion mentioned in Chapter 4-5. I should add that a frame could also include additional slots for Selectors that activate other sets of resources, so that a frame could also activate other, appropriate Ways to Think.

18. The K-line idea was first developed in Minsky, "Plain Talk About Neurodevelopmental Epistemology," found in *Proceedings of the Fifth International Joint Conference on Artificial Intelligence* (Cambridge, Mass., 1977), and Minsky, "K-lines, a Theory of Memory," *Cognitive Science* 4 (1980): 117–133; Chapter 8 of *The Society of Mind* describes more ideas about what might happen when K-lines conflict.

19. Chapter 20-1 of *The Society of Mind,* argues that even our thoughts can be ambiguous.

20. My ideas about these "micronemes" were suggested by the "microfeatures" in David L. Waltz and Jordan Pollack, "Massively Parallel Parsing," *Cognitive Science* 9, no. 1 (1985); and some earlier ideas were in Calvin N. Mooers, "Information Retrieval on Structured Content," found in *Information Theory,* edited by C. Cherry (London: Butterworths, 1956).

21. Also, several different functions could be superimposed in the same anatomical region, by using genetically distinct lines of cells that interact mainly among themselves.

22. Later Kant claims that our minds must start with some "a priori" rules like "Every change must have a cause." Today, one might interpret this as suggesting that we're born with trans-frames equipped with slots that are disposed to link to the causes of changes. At first that effect could be achieved by a simple link to whatever preceded a recent change, and in later years we could learn to refine those links.

## CHAPTER 9. THE SELF

1. Dennett goes on to point out: "Homunculi are bogeymen only if they duplicate entire the talents they are rung in to explain. If one can get a team or committee of relatively ignorant, narrow-minded, blind homunculi to produce the intelligent behavior of the whole, this is progress."

2. See http://www.theabsolute.net/minefield/witforwisdom.html.

3. Adapted from the entry on "Cold Reading" by Bertram Forer, in Robert Todd Carroll's *The Skeptic's Dictionary: A Collection of Strange Beliefs, Amusing Deceptions, and Dangerous Delusions* (New York: Wiley, 2003).

4. Nevertheless, many feelings seem to come with varied degrees of both "positive" and "negative" intensities, and this has led some psychologists to maintain that this dimension of intensity is what distinguishes emotions from other types of mental states. See Andrew Ortony, Gerald L. Clore, and Allan Collins, *The Cognitive Structure of the Emotions* (New York: Cambridge University Press, 1988), and Chapter 28 of Minsky, *The Society of Mind* (New York: Simon & Schuster, 1986).

5. See also Chapter 13-1 of Minsky, *The Society of Mind.*

6. See Chapter 23-3 of *The Society of Mind,* about "temporal blinking."

7. Roger Schank 1995 has suggested that we mainly remember things that "make sense" because our memory systems have ways to store representations that have the form of coherent stories.

8. Verse 2 of "A Light Exists in Spring," at http://www.firstscience.com/SITE/poems/dickinson3.asp.

9. Philosophers call this "the problem of qualia." There is a superb discussion of these "subjective qualities" in Daniel Dennett, "Quining Qualia," in *Consciousness in Modern Science,* edited by A. Marcel and E. Bisiach (New York: Oxford University Press, 1988).

10. In fact, a single spot of red may not be sensed as being red; in general the colors we see depend, to a large extent, on which other colors are in its neighborhood. Although we understand some of the visual resources in our brains, we still do not yet have good explanations about, for example, how we represent separate objects and their relationships.

11. See Zenon Pylyshyn, "Is Vision Continuous with Cognition?" available at http://ruccs.rutgers.edu/faculty/ZPbbs98.html. Also see Al Seckel, *Masters of Deception* (New York: Sterling Publishing, 2004).

12. Another difference between human employees and parts of a brain is that each member of a company has personal conflicts of interest. For example, each employee is hired to increase the company's profit, but this conflicts with each employee's ambition to earn more salary.

# BIBLIOGRAPHY

Acerra et al. 1999: Francesca Acerra, Yves Burnod, and Scania de Schonen. "Modeling Face Recognition Learning in Early Infant Development." In *Proceedings of European Symposium on Artificial Neural Networks,* Bruges, Belgium, April 21–23, 1999, 129–134. Text available at http://www.dice.ucl.ac.be/Proceedings/esann/esannpdf/es1999–22.pdf.

Aristotle a: Aristotle. *Nicomachean Ethics,* Book VIII, trans. W. D. Ross. Full text available at http://www.etext.library.adelaide.edu.au/a/Aristotle/nicomachean/.

Aristotle b: *Rhetoric.* Full text available at http://etext.library.adelaide.edu.au/aa8rh/.

Arnold 1865: Matthew Arnold. *Essays in Criticism.* Edited by S. R. Littlewood. London: Macmillan, 1958.

Augustine 397: Augustine. *The Confessions,* Book 10. Text available at http://www.ourladyswarriors.org/saints/augcon10.htm#chap10.

Baars 1996: Bernard J. Baars. "Understanding Subjectivity: Global Workspace Theory and the Resurrection of the Observing Self." *Journal of Consciousness Studies* 3, no. 3 (1996): 211–216. Also available at http://www.imprint.co.uk/online/baars.html.

Bacon 1620: Francis Bacon. *Novum Organum.* Available at http://etext.library.adelaide.edu.au/b/bacon/francis/organon/.

Battro 2000: Antonio M. Battro. *Half a Brain Is Enough.* Cambridge, U.K.: Cambridge University Press, 2000. Also see http://www.nobel.se/medicine/laureates/1981/sperry-lecture.html.

Blackmore 1999: Susan Blackmore. *The Meme Machine.* New York: Oxford University Press, 1999.

Bowlby 1973a: John Bowlby. *Attachment.* New York: Basic Books, 1973.

Bowlby 1973b: John Bowlby. *Separation.* New York: Basic Books, 1973.

Calvin 1966: William H. Calvin. *How Brains Think.* New York: Basic Books, 1966.

Calvin and Ojemann 1994: William H. Calvin and George A. Ojemann. *Conversations with Neil's Brain: The Neural Nature of Thought and Language.* Reading, Mass.: Addison-Wesley, 1994. Also available at http://williamcalvin.com/bk7/bk7.htm.

Carlson 1985: Shawn Carlson. "A Double-Blind Test of Astrology." *Nature* 318 (December 5, 1985): 419.

Carroll 2003: Adapted from the entry on "Cold Reading" by Bertram Forer, in Robert Todd Carroll's *The Skeptic's Dictionary: A Collection of Strange Beliefs, Amusing Deceptions, and Dangerous Delusions.* New York: Wiley, 2003. Text available at http://skepdic.com/coldread.html.

Chalmers 1995a: David J. Chalmers. "Facing Up to the Problem of Consciousness." *Journal of Consciousness Studies* 2, no. 3 (1995): 200–219. Available at http://www.u.arizona.edu/~chalmers/papers/facing.html.

Chalmers 1995b: David J. Chalmers. "The Puzzle of Conscious Experience." *Scientific American* (December 1995): 62–68.

Chandler 2004: Keith Chandler. *Australian Journal of Parapsychology* 4, no. 1 (June 2004). Also available at http://www.keithchandler.com/Essays/Savant_Syndrome.html.

Charniak 1972: Eugene Charniak. "Toward a Model of Children's Story Comprehension," PhD diss., MIT, 1972. Also available at ftp://publications.ai.mit.edu/ai-publications/pdf/AITR-266.pdf.

Clynes 1978: Manfred Clynes. *Sentics.* New York: Doubleday, 1978.

Damasio 1995: Antonio R. Damasio. *Descartes' Error.* New York: Avon Books, 1995.

Darwin 1871: Charles Darwin, *The Descent of Man.* New York: Simon and Schuster, 1986. Also available at http://www.infidels.org/library/historical/charles_darwin/descent_of_man.

Darwin 1872: Charles Darwin. *Expression of the Emotions in Man and Animals.* Edited by Paul Ekman. New York: Oxford University Press, 1998.

Davies 1992: Robertson Davies. *Tempest-Tost.* New York: Penguin, 1992.

Dawkins 1986: Richard Dawkins. *The Blind Watchmaker: Why the Evidence of Evolution Reveals a Universe Without Design.* New York: W. W. Norton, 1986.

Dawkins 1989: Richard Dawkins. *The Selfish Gene.* New York: Oxford University Press, 1989.

Dennett 1978: Daniel C. Dennett. "Why You Can't Build a Machine That Feels Pain." In *Brainstorms.* Cambridge, Mass.: MIT Press, 1978, 190–229.

Dennett 1984: Daniel C. Dennett. *Elbow Room: The Varieties of Free Will Worth Wanting.* New York: Oxford University Press, 1984.

Dennett 1988: Daniel Dennett. "Quining Qualia." In *Consciousness in Modern Science.* Edited by A. Marcel and E. Bisiach. New York: Oxford University Press, 1988. Reprinted in *Readings in Philosophy and Cognitive Science.* Edited by A. Goldman. Cambridge, Mass.: MIT Press, 1993. Text also available at http://cogprints.org/254/00/quinqual.htm.

Dennett 1991: Daniel C. Dennett. *Consciousness Explained*. Boston: Little, Brown, 1991.

Dennett and Kinsbourne 1992a: Daniel C. Dennett and Marcel Kinsbourne. "Time and the Observer." *Behavioral and Brain Sciences* 15, no. 2 (1992): 183–247. Also available at http://cogprints.ecs.soton.ac.uk/archive/00000264/.

Dennett 1992b: Daniel C. Dennett. "The Self as a Center of Narrative Gravity." In *Self and Consciousness: Multiple Perspectives*. Edited by F. Kessel, P. Cole, and D. Johnson. Hillsdale, N.J.: Erlbaum, 1992. Also available at http://cogprints.org/266/00/selfctr.htm.

Dennett 1995: Daniel C. Dennett. *Darwin's Dangerous Idea*. New York: Simon and Schuster, 1995.

Descartes 1637: Rene Descartes, in *Discours de la méthode*. "*Et le second est que, bien qu'elles fissent plusieurs choses aussi bien, ou peut-être mieux qu'aucun de nous, elles manqueraient infailliblement en quelques autres, par lesquelles on découvrirait qu'elles n'agiraient pas par connaissance, mais seulement par la disposition de leurs organes. Car, au lieu que la raison est un instrument universel, qui peut servir en toutes sortes de rencontres, ces organes ont besoin de quelque particulière disposition pour chaque action particulière; d'où vient qu'il est moralement impossible qu'il y en ait assez de divers en une machine pour la faire agir en toutes les occurrences de la vie, de même façon que notre raison nous fait agir.*"

Drescher 1991: Gary Drescher. *Made-Up Minds*. Cambridge, Mass.: MIT Press, 1991.

Egan 1998: Greg Egan. *Diaspora*. Agawan, Mass.: Millennium Press, 1998.

Einstein 1950: Albert Einstein. *Out of My Later Years*. New York: Philosophical Library, 1950, 15–20.

Evans 1963: Thomas G. Evans. "A Heuristic Program to Solve Geometric-Analogy Problems." PhD diss., MIT, 1963. Abridged version in Minsky 1968, 271–353.

Feigenbaum and Feldman 1963: Edward. A. Feigenbaum and Julian Feldman, eds. *Computers and Thought*. New York: McGraw-Hill, 1963.

Feynman 1965: Richard Feynman. *The Character of Physical Law*. Cambridge, Mass.: MIT Press, 1965, 168.

Fodor 1992: J. A. Fodor. "The Big Idea: Can There Be a Science of the Mind?" *Times Literary Supplement*, July 1992, 5–7.

Fodor 1998: Jerry Fodor. "The Trouble with Psychological Darwinism." *London Review of Books* 20, no. 2 (January 22, 1998). Also available at http://www.lrb.co.uk/v20/n02/contents.html.

Franklin 1772: Benjamin Franklin, Letter to Joseph Priestly, September 19, 1772. Text available at http://www.historycarper.com/resources/twobf3/letter11.htm.

Freud 1920: Sigmund Freud. *A General Introduction to Psychoanalysis*. New York: Boni and Liveright, 1920, 259.

Friedrick-Cofer and Huston 1986: L. Friedrick-Cofer and A. C. Huston. "Television Violence and Aggression: The Debate Continues." *Psychological Bulletin* 100 (1986): 364–371.

Gardner 2000: Howard Gardner. *Intelligence Reframed: Multiple Intelligences for the 21st Century.* New York: Basic Books, 2000.

Goodall 1968: Jane van Lawick-Goodall. "The Behavior of Free-Living Chimpanzees in the Gombe Stream Reserve." *Animal Behavior Monogram I* (1968): 161–311.

Gregory 1998: Richard Gregory. "Brainy Mind." In *British Medical Journal* 317 (1998), 1693. Also at www.richardgregory.org/papers/brainy_mind/brainy-mindhtm.

Gunkel 2006: Patrick Gunkel's ideas about ideas can be seen at http://ideonomy.mit.edu.

Haase 1986a: Kenneth W. Haase. "Exploration and Invention in Discovery." PhD diss., MIT, 1986. Text available at http://web.media.mit.edu/~haase/thesis.

Haase 1986b: Kenneth W. Haase. "Discovery Systems." In *Advances in Artificial Intelligence.* North-Holland: European Conference on Artificial Intelligence, 1986.

Haase 1987: Kenneth W. Haase. "Typical: A Knowledge Representation System for Automated Discovery and Inference." Technical Report 922. Cambridge, Mass.: MIT Artificial Intelligence Laboratory, 1987.

Haber 1979: R. N. Haber. "20 Years of Haunting Eidetic Imagery: Where's the Ghost?" *Behavioral and Brain Sciences* 2 (1979): 583–629.

Hadamard 1945: Jacques Hadamard. *The Phychology of Invention in the Mathematical Field.* New York: Dover, 1945.

Harlow 1958: Harry Harlow. "The Nature of Love." *American Psychologist* 13 (1958): 573–685. Available at http://psychclassics.yorku.ca/Harlow/love.htm.

Hayes 1997: Pat Hayes. "The Onset of Consciousness in Speech." Psyche Discussion Forum. September 29, 1997. See http://listserv.uh.edu/cgi-bin/wa?A2=ind9709&L=psyche-b&T=0&F=&S=&P=5262.

Hinde 1971: R. A. Hinde and Y. Spencer-Booth. "Towards Understanding Individual Differences in Rhesus Mother-Infant Interaction." *Animal Behaviour* 19 (1971): 165–173.

Hoffman 1996: Howard Hoffman. *Amorous Turkeys and Addicted Ducklings.* Boston: Authors Cooperative, 1996.

Horner and Gorman 1998: John R. Horner and James Gorman. *Digging Dinosaurs.* New York: Harper and Row, 1998, chapter 4.

Hume 1748: David Hume. *An Enquiry Concerning Human Understanding.* See 1777 edition at http://www.etext.leeds.ac.uk/hume/ehu/ehupbsb.htm.

Hume 1757: David Hume. *The Natural History of Religion.* Available at http://www.soci.niu.edu/~phildept/Dye/NaturalHistory.html.

James 1890: William James. *The Principles of Psychology.* New York: Simon and Schuster, 1997. Full text available at http://psychclassics.yorku.ca/James/Principles/preface.htm.

James 1902: William James. *The Varieties of Religious Experience.* New York: Random House, 1994.

Jamison 1994: Kay Redfield Jamison. *Touched with Fire: Manic-Depressive Illness and the Artistic Temperament.* New York: Free Press, 1994, 47–48.

Jamison 1995: Kay Redfield Jamison, "Manic-Depressive Illness and Creativity." *Scientific American* 272, no. 2 (February 1995): 62–67.

Johnston 1997: Elizabeth Johnston. "Infantile Amnesia." Available at http://pages.slc .edu/~ebj/IM_97/Lecture6/L6.html.

Kaiser 2006: Peter Kaiser. *The Joy of Visual Perception.* Available at http://www.yorku .ca/eye/thejoy.htm.

Kant 1787: Immanuel Kant. *Critique of Pure Reason.* London: Dent, 1991.

Koestler 1965: Arthur Koestler. *The Act of Creation.* New York: Macmillan, 1964.

Korzybski 1933: Alfred Korzybski. *Science and Sanity,* Lakeville, Conn.: International Non-Aristotelian Library, 1958. Full text at http://www.esgs.org/uk/art/sands.htm.

Laird 1987: John Laird, Allen Newell, and Paul S. Rosenbloom. "Soar: An Architecture for General Intelligence." *Artificial Intelligence* 33, no. 1 (1987). See also http://ai.eecs.umich.edu/soar/sitemaker/docs/misc/GentleIntroduction-2006.pdf.

Lakoff 1980: George Lakoff and Mark Johnson. *Metaphors We Live By.* Chicago: University of Chicago Press, 1980.

Lakoff 1992: George Lakoff. "The Contemporary Theory of Metaphor." In *Metaphor and Thought,* 2nd edition. Edited by Andrew Ortony. Cambridge, U.K.: Cambridge University Press, 1993. Also available at http://www.ac.wwu.edu/~market/semiotic/ lkof_met.html.

Landauer 1986: Thomas K. Landauer. "How Much Do People Remember?" *Cognitive Science* 10 (1986): 477–493.

Langley et al. 1987: Pat Langley, Herbert A. Simon, Gary L. Bradshaw, and Jan M. Zytkow. *Scientific Discovery: Computational Explorations of the Creative Processes.* Cambridge: MIT Press, 1987.

Lawler 1985: Robert W. Lawler. *Computer Experience and Cognitive Development: A Child's Learning in a Computer Culture.* New York: John Wiley and Sons, 1985.

Lenat 1976: Douglas B. Lenat. "AM: An Artificial Intelligence Approach to Discovery in Mathematics as Heuristic Search." PhD diss., Stanford University, 1976.

Lenat and Brown 1983a: Douglas B. Lenat and John S. Brown. "Why AM and Eurisko Appear to Work." *Artificial Intelligence* 23 (1983).

Lenat 1983b: Douglas B. Lenat. "Eurisko: A Program Which Learns New Heuristics and Domain Concepts." *Artificial Intelligence* 21 (1983). Also available at http://web .media.mit.edu/~haase/thesis/node52.html.

Lenat and Shepard 1990: Douglas B. Lenat and Mary Shepard. *CYC: Representing Encyclopedic Knowledge.* Digital Press, 1990.

Lenat 1997: Douglas Lenat. "Artificial Intelligence as Common Sense Knowledge." *Truth Journal.* Available at http://www.leaderu.com/truth/2truth07.html.

Lenat 1998: Douglas B. Lenat. *The Dimensions of Context Space*. Available at http://www.cyc.com/doc/context-space.pdf.

Lewis 1982. F. M. Lewis. "Experienced Personal Control and Quality of Life in Late Stage Cancer Patients." *Nursing Research* 31, no. 2 (1982): 113–119.

Lewis 1995a: Michael Lewis. *Shame: The Exposed Self.* New York: Free Press, 1995.

Lewis 1995b: Michael Lewis. "Self-Conscious Emotions." *American Scientist* 83 (January 1995).

Lorenz 1970: Konrad Lorenz. *Studies in Animal and Human Behaviour,* vol. 1. Translated by Robert Martin. Cambridge: Harvard University Press, 1970, 132.

Lovecraft 1926: H. P. Lovecraft. *The Call of Cthulhu and Other Weird Stories*. Edited by S. T. Joshi. New York: Penguin Books, 1999.

Luria 1968: Alexander R. Luria. *The Mind of a Mnemonist*. Cambridge, Mass.: Harvard University Press, 1968.

McCarthy 1959: John McCarthy. "Programs with Common Sense." In *Procceedings of the Symposium on Mechanization of Thought Processes,* vol. 1. Edited by D. V. Blake and A. M. Uttley. National Physical Laboratory, Teddington, England, HMSO, London, 1959, 5–27. Also available at http://www-formal.stanford.edu/jmc/mcc59.html.

McCurdy 1960: Harold G. McCurdy. "The Childhood Pattern of Genius." *Horizon Magazine,* May 1960, 32–38.

McDermott 1992: Drew McDermott. In comp.ai.philosophy. February 7, 1992.

Meltzoff and Moore 1997: Andrew N. Meltzoff and M. Keith Moore. "Explaining Facial Imitation: A Theoretical Model." *Early Development and Parenting* 6 (1997): 179–192. Also available at http://ilabs.washington.edu/meltzoff/pdf/97Meltzoff_Moore_FacialImit.pdf.

Melzack and Wall 1965: Ronald Melzack and Patrick Wall. In "Pain Mechanisms: A New Theory." *Science* 150 (1965): 975. Also see Tania Singer et al. Available at http://www.fil.ion.ucl.ac.uk/~tsinger/publications/singer_science_2004.pdf.

Melzack 1993: Ronald Melzack. "Pain: Past, Present and Future." *Canadian Journal of Experimental Psychology* 47 (1993): 615–629.

Merkle 1988: See Ralph Merkle's description at http://www.merkle.com/human Memory.html.

Minsky 1954: Marvin Minsky. "Neural-Analog Networks and the Brain-Model Problem." PhD diss., Princeton University, 1954.

Minsky 1956: M. L. Minsky. "Heuristic Aspects of the Artificial Intelligence Problem." Lincoln Laboratory, MIT. Group Rept. 34–55, ASTIA Doc. No. 236885, December 1956.

Minsky 1968: Marvin Minsky, ed. *Semantic Information Processing.* Cambridge, Mass.: MIT Press, 1968. This anthology is currently out of print, but my chapter,

"Matter, Mind, and Models," is also available at http://web.media.mit.edu/~minsky/ papers/MatterMindModels.html.

Minsky and Papert 1971: Marvin Minsky and Seymour Papert. *Progress Report on Artificial Intelligence.* Available at http://web.media.mit.edu/~minsky/papers/ PR1971.html.

Minsky 1974: Marvin Minsky. *A Framework for Representing Knowledge.* Cambridge, Mass.: MIT Press, 1974. Also available at http://web.media.mit.edu/~minsky/papers/ Frames/frames.html.

Minsky 1977: Marvin Minsky. "Plain Talk About Neurodevelopmental Epistemology." In *Proceedings of the Fifth International Joint Conference on Artificial Intelligence,* Cambridge, Mass., 1977.

Minsky 1980a: Marvin Minsky. "Jokes and Their Relation to the Cognitive Unconscious." In *Cognitive Constraints on Communication.* Edited by Lucia Vaina and Jaakko Hintikka. Boston: Reidel, 1981. Also available at web.media.mit.edu/~minsky/ papers/jokes.cognitive.txt.

Minsky 1980b: Marvin Minsky. "K-lines, a Theory of Memory." *Cognitive Science* 4 (1980): 117–133.

Minsky 1981: Marvin Minsky. "Music, Mind, and Meaning." *Computer Music Journal* 5, no. 3 (Fall 1981). Also available at web.media.mit.edu/~minsky/papers/ MusicMindMeaning.html.

Minsky 1986: Marvin Minsky. *The Society of Mind.* New York: Simon and Schuster, 1986.

Minsky and Papert 1988: Marvin Minsky and Seymour Papert. *Perceptrons,* 2nd edition. Cambridge: MIT Press, 1988.

Minsky 1991: Marvin Minsky. "Logical vs. Analogical or Symbolic vs. Connectionist or Neat vs. Scruffy." In *Artificial Intelligence at MIT: Expanding Frontiers,* vol. 1. Edited by Patrick H. Winston. Cambridge: MIT Press, 1990. Also available at http:// web.media.mit.edu/~minsky/papers/SymbolicVs.Connectionist.html.

Minsky 1992: Marvin Minsky. "Future of AI Technology." *Toshiba Review* 47, no. 7 (July 1992). Also available at http://web.media.mit.edu/~minsky/papers/Causal Diversity.html.

Minsky et al. 2004: Marvin Minsky, Push Singh, and Aaron Sloman. "The St. Thomas Common Sense Symposium: Designing Architectures for Human-Level Intelligence." *AI Magazine,* Summer 2004.

Minsky 2005: Marvin Minsky. "Interior Grounding, Reflection, and Self-Consciousness." In *Proceedings of an International Conference on Brain, Mind and Society,* Tohoku University, Japan, September 2005. See http://www.ic.is.tohoku.ac.jp/~GSIS/. Also available at http://web.media.mit.edu/~minsky/papers/Internal%20Grounding.html.

Mooers 1956: Calvin N. Mooers. "Information Retrieval on Structured Content." In *Information Theory.* Edited by C. Cherry. London: Butterworths, 1956.

Mueller 2006: Erik T. Mueller. *Commonsense Reasoning*. Boston: Elsevier Morgan Kaufmann, 2006.

Nelson 2001: Charles A. Nelson. "The Development and Neural Bases of Face Recognition." *Infant and Child Development* 10 (2001): 3–18. Also available at http://www.biac.duke.edu/education/courses/spring03/cogdev/readings/C.A.%20Nelson%20(2001).pdf.

Newell 1955: Allen Newell. "The Chess Machine." In *Proceedings of Western Joint Computer Conference,* March 1955.

Newell et al. 1960a: A. Newell, J. C. Shaw, and H. A. Simon. "Report on a General Problem Solving Program." In *Proceedings of the International Conference on Information Processing.* Paris: UNESCO, 1960, 256–264.

Newell et al. 1960b: Allen Newell, J. Clifford Shaw, and Herbert A. Simon. "A Variety of Intelligent Learning in a General Problem Solver." In *Self-Organizing Systems.* Edited by M. T. Yovitts and S. Cameron. New York: Pergamon Press, 1960. I have seen few references to this profound idea since its original publication!

Newell and Simon 1963: Allen Newell and Herbert A. Simon. "GPS, a Program That Simulates Human Thought." In *Computers and Thought.* Edited by E. A. Feigenbaum and J. Feldman. New York: McGraw-Hill, 1963.

Newell 1972: Allen Newell and Herbert A. Simon. *Human Problem Solving.* Englewood Cliffs, N.J.: Prentice Hall, 1972. See also http://sitemaker.umich.edu/soar for a description of the problem-solving architecture called *SOAR.*

O'Regan 2006: See Kevin O'Regan's articles on "Change-Blindness" at http://nivea.psycho.univ-paris5.fr/ECS/ECS-CB.html and "Change Blindness as a Result of Mudsplashes" in *Nature,* August 2, 1998.

Ortony et al. 1988: Andrew Ortony, Gerald L. Clore, and Allan Collins. *The Cognitive Structure of the Emotions.* New York: Cambridge University Press, 1988.

Osterweis et al. 1987: Marian Osterweis, Arthur Kleinman, and David Mechanic. *Pain and Disability: Clinical, Behavioral, and Public Policy Perspectives.* Washington, D.C.: National Academy Press, 1987.

Pagels 1988: Heinz Pagels. *The Dreams of Reason: The Computer and the Rise of the Sciences of Complexity.* New York: Simon and Schuster, 1988.

Panksepp 1988. Jaak Panksepp and Manfred Clynes, eds. *Emotions and Psychopathology.* New York: Plenum Press, 1988.

Panksepp 1998: Jaak Panksepp. *Affective Neuroscience.* New York: Oxford University Press, 1998.

Pepperberg 1998: See reports on Irene Pepperberg's research on parrots at http://www.alexfoundation.org/irene.htm. Also available at http://pubpages.unh.edu/~jel/video/alex.html.

Perkins 1981: David N. Perkins. *The Mind's Best Work.* Cambridge, Mass.: Harvard University Press, 1981.

Phillips 2004: Melissa Lee Phillips. *Seeing with New Sight.* Available at http://faculty .washington.edu/chudler/visblind.html.

Piaget 1923: Jean Piaget. *The Language and Thought of the Child.* New York: Routledge, 2001.

Pivar and Finkelstein 1966: M. Pivar and M. Finkelstein. *The Programming Language LISP: Its Operation and Applications.* Cambridge, Mass.: MIT Press, 1966. Text available at http://community.computerhistory.org/scc/projects/LISP/book/ III_LispBook_Apr66.pdf#page=270.

Plsek 1996: Paul E. Plsek. *Models for the Creative Process.* Available at www.directed creativity.com/pages/WPModels.html.

Pohl 1970: Frederik Pohl. *Day Million.* New York: Ballantine Books, 1970.

Poincaré 1913: Henri Poincaré. *The Foundations of Science, Science and Hypothesis: The Value of Science, Science and Method,* New York: Science Press, 1929.

Pólya 1954: George Pólya. *Induction and Analogy in Mathematics.* Princeton: Princeton University Press, 1954.

Pólya 1962: George Pólya. *Mathematical Discovery: On Understanding, Learning and Problem Solving.* New York: Wiley, 1962.

Proust 1927: Marcel Proust. *Remembrance of Things Past.* New York: Random House, 1927–1932.

Pylyshyn 1998: Zenon Pylyshyn. "Is Vision Continuous with Cognition." Available at http://ruccs.rutgers.edu/faculty/ZPbbs98.html.

Quillian 1966: Ross Quillian. "Semantic Memory." PhD diss., Carnegie Institute of Technology, 1966. Reprinted in Minsky 1968.

Ramachandran 2004: V. S. "Ramachandran." "Beauty or Brains?" *Science* 305, no. 5685 (August 6, 2004).

Rosenfeld 1996: Ronald Rosenfeld. "A Maximum Entropy Approach to Adaptive Statistical Language Modeling." *Computer, Speech and Language* 10 (1996). Also available at http://www.cs.cmu.edu/afs/cs/user/roni/WWW/me-csl-revised.ps.

Royce 1908: Josiah Royce. *The Philosophy of Loyalty.* Nashville: Vanderbilt University Press, 1995.

Ryle 1949: Gilbert Ryle. *The Concept of Mind.* Chicago: University of Chicago Press, 1949.

Samuel 1959: Arthur L. Samuel. "Some Studies in Machine Learning Using the Game of Checkers." *IBM Journal of Research and Development* 3 (July 1959): 211–219.

Schaffer and Emerson 1964: H. R. Schaffer and P. E. Emerson. "The Development of Social Attachments in Infancy." *Monographs for the Society of Research in Child Development* 29, no. 3 (1964).

Schank 1975: Roger C. Schank. *Conceptual Information Processing.* New York: American Elsevier, 1975.

Schank and Abelson 1977: Roger Schank and Robert Abelson. *Scripts, Goals, Plans and Understanding*. Mahwah, N.J.: Erlbaum Associates, 1977.

Schank 1995: Roger Schank. *Tell Me a Story*. Evanston, Ill.: Northwestern University Press, 1995.

Seay et al. 1964: B. Seay, B. R. Alexander, and H. F. Harlow. "Maternal Behavior of Socially Deprived Rhesus Monkeys." *Journal of Abnormal and Social Psychology* 69, no. 4 (1964): 345–354.

Seckel 2004: Al Seckel. *Masters of Deception*. New York: Sterling Publishing, 2004.

Shannon 1948: Claude E. Shannon. "A Mathematical Theory of Communication." *Bell System Technical Journal* 27 (July and October 1948). Also available at http://cm.bell-labs.com/cm/ms/what/shannonday/shannon1948.pdf.

Singh et al. 2002: Push Singh, Thomas Lin, Erik T. Mueller, Grace Lim, Travell Perkins, and Wan Li Zhu. "Open Mind Common Sense: Knowledge Acquisition from the General Public." *Proceedings of the First International Conference on Ontologies, Databases, and Applications of Semantics for Large Scale Information Systems*, Irvine, Calif.

Singh 2003a: Push Singh. "Examining the Society of Mind." *Computing and Informatics* 22, no. 5 (2003): 521–543. This article briefly describes the history of the Society of Mind theory, explains some of its essential components, and relates it to recent developments in Artificial Intelligence. Also available at http://web.media.mit.edu/~push/ExaminingSOM.pdf.

Singh and Minsky 2003b: Push Singh and Marvin Minsky. "An Architecture for Combining Ways to Think." In *Proceedings of the International Conference on Knowledge Intensive Multi-Agent Systems*, Cambridge, Mass.

Singh 2003c: "A Preliminary Collection of Reflective Critics for Layered Agent Architectures." In *Proceedings of the Safe Agents Workshop*, Melbourne, Australia: AAMAS, 2003. Describes a class of agents concerned with noticing recent problems in the deliberations of a layered commonsense reasoning system. Also available at http://web.media.mit.edu/~push/ReflectiveCritics.pdf.

Singh et al. 2004: Push Singh, Marvin Minsky, and Ian Eslick. "Computing Commonsense." *BT Technology Journal* 22, no. 4 (October 2004). Also available at http://web.media.mit.edu/~push/Computing-Commonsense-BTTJ.pdf.

Singh 2005a: Push Singh. *EM-ONE: An Architecture for Reflective Commonsense Thinking*, PhD thesis, MIT, June 2005. Also available at http://web.media.mit.edu/~push/push-thesis.pdf.

Singh and Minsky 2005b: Push Singh and Marvin Minsky. "An Architecture for Cognitive Diversity." In *Visions of Mind*. Edited by Darryl Davis. London: Idea Group, 2005.

Sloman 1992: Aaron Sloman. "Developing Concepts of Consciousness." *Behavioral and Brain Sciences* 14 (1992).

Solomonoff 1957: Raymond J. Solomonoff. "An Inductive Inference Machine." *IRE Convention Record.* Section on Information Theory, Part 2 (1957): 56–62.

Solomonoff 1964: R. J. Solomonoff. "A Formal Theory of Inductive Inference." *Information and Control* 7 (1964): 1–22.

Solomonoff 1997: R. J. Solomonoff. "The Discovery of Algorithmic Probability." *Journal of Computer and System Sciences* 55, no. 1 (1997). Also available at http://world.std.com/~rjs/barc97.html.

Spencer-Booth and Hinde 1971: Y. Spencer-Booth and R. A. Hinde. *Animal Behavior* 19 (1971): 174–191, 595–605.

Spencer-Brown 1972: G. Spencer-Brown. *Laws of Form.* New York: Crown Publishers, 1972.

Sri Chinmoy 2003: Sri Chinmoy. "Consciousness." Text available at http://www.yoga ofsrichinmoy.com/the_higher_worlds/consciousnesss/.

Stickgold et al. 2000: Robert Stickgold, April Malia, Denise Maguire, David Roddenberry, Margaret O'Connor. "Replaying the Game: Hypnagogic Images in Normals and Amnesics." *Science* 290, no. 5490 (October 13, 2000): 350–353.

Thagard 2001: Paul Thagard. "How to Make Decisions: Coherence, Emotion, and Practical Inference." In *Varieties of Practical Inference.* Edited by E. Millgram. Cambridge, Mass.: MIT Press, 2001. Text also available at http://cogsci.uwaterloo .ca/Articles/Pages/how-to-decide.html.

Thorndike 1911: Edward L. Thorndike. *Animal Intelligence.* New York: Macmillan, 1911, 244.

Tinbergen 1951: Nikolaas Tinbergen. *The Study of Instinct.* London: Oxford University Press, 1951.

Turing 1936: Alan Turing. "On Computable Numbers." Available at http://www .abelard.org/turpap2/tp2-ie.asp#section-1.

Turing 1950: Alan Turing. "Computing Machinery and Intelligence." *Mind* 49 (1950). Also available at http://cogprints.org/499/00/turing.html and at www.cs.swarthmore .edu/~dylan/Turing.html.

Viezzer 2000: Manuela Viezzer. "Ontologies and Problem-Solving Methods." 14th European Conference on Artificial Intelligence, Humboldt University, Berlin, August 2000. Also available at www.cs.bham.ac.uk/~mxv/publications/onto_engineering.

Vinacke 1952: W. E. Vinacke. *The Psychology of Thinking.* New York: McGraw-Hill, 1952.

Waltz and Pollack 1985: David L. Waltz and Jordan Pollack. "Massively Parallel Parsing." *Cognitive Science* 9, no. 1 (1985).

Watts 1960: Alan Watts. *This Is It.* New York: Random House, 1960.

Wertheimer 1945: Max Wertheimer. *Productive Thinking.* New York: Harper and Brothers, 1945.

West 1928: Rebecca West. *The Strange Necessity,* New York: Doubleday, 1928.

Wilde 1905: Oscar Wilde. *De Profundis.* New York: Knickerbocker Press, 1905. Text at http://www.gutenberg.org/dirs/etext97/dprof10.txt.

Winston 1970: Patrick Winston. "Learning Structural Descriptions from Examples." PhD diss. MIT, 1970. Also available at https://dspace.mit.edu/bitstream/1721.1/6884/2/AITR-231.pdf.

Winston 1975: Patrick H. Winston, ed. *The Psychology of Computer Vision.* New York: McGraw-Hill, 1975.

Winston 1984: Patrick H. Winston. *Artificial Intelligence,* 3rd edition. Boston: Addison-Wesley, 1984, 1992. (The 1984 first edition is easier for beginners.)

Wundt 1897: Wilhelm Wundt. *Outlines of Psychology.* Translated by C. H. Judd. Full text available at http://psychclassics.yorku.ca/Wundt/Outlines/.

# INDEX

A-Brain, 100–102, 186, 196, 197
Absolutism, 95
Acerra, Francesca, 57
Actions, Trans-Frames to
  represent, 283–84
Adaptive skills, 174
Adult emotions, 27–30
Adventurousness, 278, 324
*Aesop's Fables,* 58–59
Affection, 37, 44
Aims, 33, 188
Alarm, 44
Alarms, built-in, 229, 338
Allen, Woody, 70, 76
Ambiguities, 169
*American, The* (James), 89
Amnesia of Infancy, 182
Analogies, 124, 271, 343–44
  Difference-Networks in
    construction of, 202
  parallel, *see* Panalogies
  reasoning by, 205–9, 226–27
  space-related, 125
Anderson, Poul, 220
Anger, 3, 4, 12, 13, 23
  behavior and, 34

embodiment of, 233, 234
pain and, 72
protective function of, 24
self-induced, 91–92
switching on, 93
Anguish, 18, 69, 70
Animals, attachments of, 53–60
Anxiety, 37, 44, 231
*Arabian Nights, The,* 202
Archetypes, 310
Aristotle, 192, 202, 234, 296
  on differences, 187
  on grief, 80
  on pleasure vs. pain, 321
  on shame, 40
Arnold, Matthew, 235, 237
Artificial Intelligence (AI), 1, 6,
  139, 226
  "baby-machine" approach to,
    178–82, 290
  credit assignment in, 358$n11$
  mental bugs in, 341
  problem solving in, 206–9
Art, 198
Asimov, Isaac, 42
Asking for help, 228